"This is an erudite book by one of the leading Catholic philosophers of our time. Bracken shows that if God plays dice, God does not play alone."

—Ilia Delio, OSF
Woodstock Theological Center
Georgetown University

"Weaving together the wisdom of Whitehead, issues in ancient philosophy and contemporary science, and his commitment to Christian theology, Joseph Bracken offers compelling arguments that affirm God works providentially in the world. Bracken finely tunes and ardently articulates his model of a Triune God who empowers creatures in a world of freedom and chance. Of his many books, I find this one most lucid, as Bracken takes his well-percolated insights and applies them to concrete issues in theology, philosophy, and science."

—Thomas Jay Oord
Professor of Theology and Philosophy
Northwest Nazarene University, Nampa, Idaho
Author of *Defining Love: A Philosophical, Scientific and Theological Engagement*

"Bracken's book will appeal to thoughtful Christians convinced of the need for a radical transformation of theological discourse in a post-mechanistic era. This intellectually challenging work sums up the author's many years of rethinking Christian faith within the framework of Alfred North Whitehead's philosophy."

—John F. Haught, PhD
Woodstock Theological Center
Georgetown University

Does God Roll Dice?

Divine Providence for a World in the Making

Joseph A. Bracken, SJ

A Michael Glazier Book

LITURGICAL PRESS
Collegeville, Minnesota

www.litpress.org

A Michael Glazier Book published by Liturgical Press

Cover design by Ann Blattner. Photo: Thinkstock/istockphoto.

Scripture texts in this work are taken from the *New American Bible with Revised New Testament and Revised Psalms* © 1991, 1986, 1970 Confraternity of Christian Doctrine, Washington, DC, and are used by permission of the copyright owner. All Rights Reserved. No part of the *New American Bible* may be reproduced in any form without permission in writing from the copyright owner.

Excerpts reprinted with the permission of Scribner, a Division of Simon & Schuster, Inc., from PROCESS AND REALITY by Alfred North Whitehead. Copyright © 1929 by The MacMillan Company; copyright renewed 1957 by Evelyn Whitehead. Copyright © 1978 by Free Press, a Division of Simon & Schuster. All rights reserved.

© 2012 by Order of Saint Benedict, Collegeville, Minnesota. All rights reserved. No part of this book may be reproduced in any form, by print, microfilm, microfiche, mechanical recording, photocopying, translation, or by any other means, known or yet unknown, for any purpose except brief quotations in reviews, without the previous written permission of Liturgical Press, Saint John's Abbey, PO Box 7500, Collegeville, Minnesota 56321-7500. Printed in the United States of America.

Library of Congress Cataloging-in-Publication Data

Bracken, Joseph A.
　　Does God roll dice? : divine providence for a world in the making / Joseph Bracken.
　　　　p.　　cm.
　　"A Michael Glazier book."
　　Includes bibliographical references and index.
　　ISBN 978-0-8146-8052-0 — ISBN 978-0-8146-8053-7 (e-book)
　　1. Causation. 2. Cosmology. 3. One (The One in philosophy) 4. Many (Philosophy) 5. Philosophical theology. 6. Providence and government of God. I. Title.

BD591.B73 2012
230—dc23   2011042741

Contents

Acknowledgments   vii

Introduction   ix

Part One: Philosophical Cosmology / Natural Theology from an Evolutionary Perspective

Chapter One: Is There a Reason for Everything, or Do Some Things Just Happen?   3

Chapter Two: Value and Creativity   15

Chapter Three: Intelligent Design and the Self-Organization of Nature   31

Chapter Four: Rethinking Primary and Secondary Causality   47

Part Two: Systems Thinking in the Social Sciences

Chapter Five: From Platonic Forms to Open-Ended Systems: The Search for Truth and Objectivity   61

Chapter Six: Whiteheadian Societies as Self-Unifying Systems   75

Chapter Seven: Subjectivity and Objectivity within Open-Ended Systems   91

Chapter Eight: The Democratic Process as an Open-Ended System in Political Life   111

Part Three: Christian Doctrinal Questions

Chapter Nine: Incarnation and Redemption within the Cosmic Process   127

Chapter Ten: A New Look at the Resurrection of the Body 139

Chapter Eleven: Church and Sacraments from a Process Perspective 155

Chapter Twelve: Inclusivity and Exclusivity in a Religious Context 167

Conclusion 177

Selected Bibliography 185

Index 191

Acknowledgments

In this book frequent references are made to the works of Alfred North Whitehead, particularly to his most systematic work, *Process and Reality*, in the corrected edition edited by David Ray Griffin and Donald W. Sherburne and published by the Free Press in New York City in 1978. I am grateful to Simon and Schuster, Inc., for permission to cite material from this source. Likewise, I am grateful to the Templeton Foundation Press for permission to use material from two of my previous publications under their auspices: *Christianity and Process Thought: Spirituality for a Changing World* (2006) and *Subjectivity, Objectivity, and Intersubjectivity: A New Paradigm for Religion and Science* (2009). In many ways, these two books together with the present book supplement one another. A lengthier version of chapter 11 of this book will be included as a chapter in a forthcoming book to be published by Wipf and Stock in Eugene, Oregon, under the general editorship of John B. Cobb Jr. with the intention of showing the pertinence of Whitehead's metaphysics to various world religions. Likewise, I draw material from my previously published articles in three books on the process philosophy of Whitehead published by Ontos Verlag in Frankfurt, Germany: "Subjectivity, System, and Intersubjectivity," in *Subjectivity, Process, and Rationality*, edited by Michel Weber and Pierfrancisco Basile (2007); "Whitehead's Rethinking of the Problem of Evil," in *Handbook of Whiteheadian Process Thought*, vol. 1, edited by Michel Weber and Will Desmond (2008); and "Systems Thinking and Emergence," in *Applied Process Thought II: Following a Trail Ablaze*, edited by Mark Dibben and Rebecca Newton (2009). In each case, of course, the material from these previous publications has been incorporated into a different context and used for a somewhat different purpose in the present book. The comparison between the philosophies of

Charles Sanders Peirce and Whitehead as laid out in chapter 1 of this book is explained in greater detail in an article titled "Feeling Our Way Forward: Continuity and Discontinuity within the Cosmic Process" in *Theology and Science* 8, no. 3 (August 2010). Some ideas in chapter 9 of this book are also to be found in an article on trinitarian Christology titled "Trinitarian Spirit Christology: In Need of a New Metaphysics" in *Theological Studies* 72, no. 4 (December 2011). Finally, the basic notion of the relative equivalence of Whiteheadian structured societies and self-unifying systems in the natural and social sciences is analyzed from a purely philosophical perspective in a forthcoming issue of *Process Studies* and in a future online book on metaphysics published by InTech (www.intechweb.org).

Introduction

"God does not play dice." This oft-quoted statement in religion and science is actually a paraphrase of what Albert Einstein wrote in a letter to fellow physicist Max Born on December 4, 1926: "Quantum mechanics is certainly imposing. But an inner voice tells me that it is not yet the real thing. The theory says a lot, but does not bring us any closer to the secret of the 'old one.' I, at any rate, am convinced that *He* [God] is not playing at dice."[1] Einstein's opposition to quantum mechanics, with its clear implication of the all-pervasive possibility of chance or contingency in the workings of nature, makes clear, as Heinz Pagels points out in *The Cosmic Code*, that he was "the last classical physicist."[2] That is, he was still convinced of the validity of Newtonian mechanics, with its deterministic laws of cause and effect, even though he helped lay the theoretical foundation of quantum theory by stipulating that light is transmitted by discrete particles, photons, rather than in successive energy waves.[3] For in Einstein's view everything in the universe is to be understood as the necessary result of a network of causes and effects.[4] Hence, radical contingency such as that demanded by quantum theory is ultimately illusory; in due time a deterministic theory based on strict cause and effect relations will eventually explain the alleged contingency of events at the subatomic level in nature.

---

1. Albert Einstein to Max Born, December 4, 1926, in *The Born-Einstein Letters 1916–1955: Friendship, Politics and Physics in Uncertain Times*, trans. Irene Born (1971; repr., Basingstoke: Macmillan, 2005), 88. Italics in original.
2. Heinz R. Pagels, *The Cosmic Code: Quantum Physics as the Language of Nature* (New York: Bantam Books, 1984), 3–23.
3. Ibid., 15.
4. Ibid., 23.

The subsequent history of quantum physics in the twentieth century, however, has called into question the theories of Einstein and others that there are "hidden variables" that explain the apparent contingency in the activity of subatomic particles. Instead, scientists in both the natural and social sciences have turned for the most part to probability theory to explain how nature works. But this shift in thinking about the laws of nature has left many Christian philosophers and theologians "holding the bag," so to speak. In the early modern period of Western civilization, their predecessors in the academy adapted the biblical understanding of the God-world relationship to the Newtonian worldview through natural theology, a philosophical/ theological theory based on the supposition that the universe is a cosmic machine and that God is both the original architect of this cosmic process and afterwards its celestial "mechanic" as needed.[5] This allowed Christian philosophers and theologians of that time to affirm on scientific grounds the biblical belief in God as Creator and Sustainer of the universe. But there was a risk in this *rapprochement* between reason and revelation. For when and if the scientific understanding of the cosmic process changed, then Christian philosophers and theologians would likewise have to change their interpretation of the God-world relationship or face the charge that Christian philosophy and theology is based on myth rather than on scientifically established fact.

Precisely such a shift in thinking about the nature of the cosmic process has happened in the twentieth century, and it has not been easy for many, if not most, Christian philosophers and theologians to accept. For example, if God does indeed play dice in guiding the universe to its appointed end, how is this to be reconciled with the traditional understanding of divine omniscience and omnipotence? For that matter, how does the biblical understanding of the end of the world (the Last Judgment, subjective immortality, and the reconciliation of all things in Christ) come to terms with current scientific projections for the end of the cosmic process in terms of either a "heat death" or a "deep freeze"? Presumably the rival claims of science and religion about the way the world works will be reconciled with one another. But in the meantime both patience and openness to new ideas seem to be needed to move beyond the current status quo. As Granville Henry points out in *Christianity and the Images of Science*, "conflict between science and religion occurs when religion, after accepting science

---

5. See John Hedley Brooke, "Natural Theology," in *Science and Religion: A Historical Introduction*, ed. Gary B. Ferngren (Baltimore, MD: Johns Hopkins University Press, 2002), 163–75.

into its theology, engages a new and different science when it arises."[6] Both sides in the religion and science debate, of course, have to be willing to modify their previous assumptions if the exchange is to bear fruit. However, at least in the early stages of this process, Christian philosophers and theologians may seem to be making more concessions to the other side than their dialogue partners in the natural and social sciences.

Keeping the present tension between the truth claims of both religion and science in key areas of human life, I set forth in this book a new look at a previously settled methodology for conducting research and reflection in both the humanities and the sciences. What I have in mind, therefore, is not a metaphysical system in the classical sense, an alleged "theory of everything." For within academic circles in Western Europe and North and South America, there is currently widespread skepticism about the possibility of a comprehensive worldview suitable for serious work not only in the humanities but even more so in the natural and social sciences. Some of this movement away from the notion of a unifying worldview has come about through an explosion of subdisciplines within originally more generic disciplines, along with specialized methodologies for doing work in those new areas of research. But likewise involved in this move away from classical metaphysics is a deep suspicion of what Emmanuel Levinas in his book *Totality and Infinity* called "totalizing systems"[7] or what postmodernists like Jean-Francois Lyotard have termed "meta-narratives,"[8] speculative theories designed to account for the meaning and value of human life and indeed of the cosmic process as a whole. As they see it, such schemes are thinly disguised attempts to promote the goals and values of particular interest groups within society to the long-term disadvantage of other groups.

Rather, my intent in this book is to set forth what might be called a heuristic structure or organizing principle for the interpretation of human experience that seems to be already generally operative in the overall methodology of the natural and social sciences and, to a lesser extent, even in the humanities. Its historical antecedents are to be found in the classical paradigm for the relation between the One and the Many, whereby the One

---

6. Granville C. Henry, *Christianity and the Images of Science* (Macon, GA: Smyth & Helwys, 1998), 28.

7. Emmanuel Levinas, *Totality and Infinity: An Essay on Exteriority*, trans. Alphonso Lingis (Pittsburgh, PA: Duquesne University Press, 1969), 35–40.

8. Jean-Francois Lyotard, *The Postmodern Condition: A Report on Knowledge*, trans. Geoff Bennington and Brian Massumi (Minneapolis, MN: University of Minnesota Press, 1984), xxiv.

is seen as a transcendent entity that gives order and intelligibility to the Many, understood as all the other entities within its own field of activity. One finds this classical understanding of the relation between the One and the Many implicitly at work in Plato's doctrine of transcendent forms as the organizing principles for all the contingent events within human sense experience. But the same basic line of thought is continued in the substance-oriented philosophy of Aristotle and in the way that Thomas Aquinas and other medieval thinkers conceived the God-world relationship.

My proposed revision of this classical understanding of the relation between the One and the Many is to the effect that the One is not an intelligible entity transcendent of the empirical Many, but rather a structured field of activity or law-like environment for the ongoing interplay of the Many with one another over an extended period of time. It is accordingly a more evolutionary understanding of the relation between the One and the Many than is presupposed in the earlier paradigm. For on the one hand, the One as a structured field of activity or law-like environment for the Many as its component parts or members serves as a relatively permanent frame of reference for the way in which the Many relate to one another in an ever-changing external environment. Yet on the other hand, the One itself evolves in its internal organization and structure as a result of the way in which it is linked to the ongoing interaction of the Many both with one another and with the external environment. Thus understood, this new paradigm for the relation of the One and the Many could be a possible starting point for explanation of the phenomenon of emergence in the natural sciences, above all in the life sciences. As Stuart Kauffman explains with his notion of "unifying systems" in chapter three of this book, a living cell as a higher-order field of activity with new patterns of organization and structure could quite unexpectedly emerge out of the interplay of interrelated subfields of activity at the molecular level. Yet even if this proves to be the case, there is thus far no fully satisfactory philosophical explanation for how this is possible, namely, how an effect can be greater than its antecedent cause.[9] Likewise, this new paradigm for the relation between the One and the Many might provide stronger

---

9. See Philip Clayton, *Mind and Emergence: From Quantum to Consciousness* (New York: Oxford University Press, 2004), esp. 31–33, 203–6. Clayton argues persuasively for a theory of strong emergence as the middle ground between physicalism and dualism in the debate over the *rationale* for evolution in the cosmic process. But in my judgment, he does not further ground his notion of strong emergence in a consistent metaphysical scheme that includes both life and nonlife as manifestations of a common metaphysical principle.

theoretical justification for the increasing acceptance of systems thinking in the contemporary social sciences, as I shall make clear in part two of this book. Finally, within the humanities one has to account for the unexpected emergence of a new metaphor or novel point of convergence between two areas of academic research that were once considered closed to one another but now in virtue of this metaphor are closely interrelated, effectively creating a new joint field of research and reflection.[10] But why and how did it happen? Could it happen again?

A few years ago I tried to vindicate this revision of the classical paradigm for the relation between the One and Many with a book grounded in a broad overview of the history of Western philosophy.[11] The key idea of the book was to highlight the different ways in which philosophers through the ages conceived the logical relation between the One and the Many, albeit without too much success. Did they, for example, give priority to the Many, their experience of the concrete differences between people and things at any given moment, or were they primarily interested in the One as a transcendent reality above and beyond the Many, which thus serves as their necessary principle of order and intelligibility? In the end I concluded that it was a mistake to prioritize the Many over the One or vice versa. The dominance of the Many over the One ultimately leads to near chaos in one's understanding of how the world works. Yet an overly strong emphasis on the One as opposed to the Many leads to suppression of legitimate differences, that is, differences that ultimately make a difference. As a result, I opted for an intrinsic interdependence of the One on the Many and the Many on the One such as can be found in a modest revision of the category of "society" in the metaphysics of Alfred North Whitehead. Furthermore, the basic premise of Whitehead's overall metaphysical scheme that "the final real things of which the world is made up" are momentary self-constituting subjects of experience (actual

---

10. See Mary Gerhart and Allan Melvin Russell, *New Maps for Old: Explorations in Science and Religion* (New York: Continuum, 2001), 31–43. Their understanding of "metaphoric process," which involves the deliberate superposition and thus conscious distortion of two previously separate "fields" of meaning so as to create a new "world" of meaning, seems to have considerable affinity with my own understanding of the relation between the One and the Many in terms of an ascending hierarchy of ever-more complex structured fields of activity both in human cognition and in the extramental physical world.

11. See Joseph A. Bracken, *Subjectivity, Objectivity and Intersubjectivity: A New Paradigm for Religion and Science* (West Conshohocken, PA: Templeton Foundation Press, 2009).

entities)[12] provided me with a key insight into how the Many in each case relate both to one another and to their external environment so as in the end to subtly reconfigure the prevailing organization and structure of the One as their common environment.

My thesis seemed to make good sense and offered a new vantage point for interpreting the ever-changing viewpoints on the structure of reality within Western philosophy. But there was an unforeseen shortcoming in the book. Only in the final chapters of the book did I apply this new paradigm for the relation between the One and the Many to various controversial issues in the field of theology and science. So in the present volume I try to remedy that shortcoming. In twelve relatively short chapters I employ what can be called a field- or context-oriented understanding of the relationship between the One and the Many to a number of specific topics under three broad headings: Philosophical Cosmology / Natural Theology (part one), Systems Thinking in the Social Sciences (part two), and Doctrinal Issues in Contemporary Christian Theology (part three). With four chapters within each of these parts, I hope to make clear how a change in one's basic understanding of the relationship between the One and the Many makes a significant difference in dealing with controversial issues in the ongoing exchange between Christian philosophers and theologians on the one side, and on the other side scientists interested in the broader humanistic implications of their own disciplines for life in this world.

Is this perhaps too much to undertake in a single, relatively short book? Does one not risk the charge of superficiality in dealing with such a broad range of topics? In response, I note that I have already laid out the main lines of my evolutionary understanding of the relation between the One and the Many in earlier publications.[13] So I am not starting from scratch in writing the present volume. Likewise, to avoid even the appearance of setting forth a totalizing system or a meta-narrative in this book, I have worked hard at a more inductive rather than a strictly deductive approach in setting forth how I deal with various controversial issues in the fields of religion and science. That is, in each chapter I first set forth a controverted issue in a given field, then briefly indicate how it has been handled

---

12. Alfred North Whitehead, *Process and Reality: An Essay in Cosmology*, corrected edition, ed. David Ray Griffin and Donald W. Sherburne (New York: Free Press, 1978), 18.

13. Cf. esp. Joseph A. Bracken, *The One in the Many: A Contemporary Reconstruction of the God-World Relation* (Grand Rapids, MI: Eerdmans, 2001), and its popularization in my subsequent book *Christianity and Process Thought: Spirituality for a Changing World* (Philadelphia, PA: Templeton Foundation Press, 2006).

by others, and finally explain how a suitably revised understanding of a Whiteheadian society would help one to deal with that same issue with a greater hope of success. In some ways, this approach is akin to what the prominent social scientist Niklas Luhmann calls a "supertheory," a heuristic structure or set of broad operational principles that can be employed in a variety of contexts to further organize and systematize the empirical data.[14] In any event, proceeding in this more inductive fashion, I may better keep the attention of the reader as I move from chapter to chapter. A professional in each of the problem areas treated in this book may well complain that I have oversimplified a much more complicated issue. But the primary purpose of the book is give the nonspecialist (e.g., the members of an adult discussion group, college undergraduates, or any other group of intelligent people) the confidence that an objective understanding of self, others, the world of nature, and (within limits) God is still possible even in this somewhat cynical postmodern era.

Finally, to return to the provocative title of this book, the answer to the question "Does God roll dice?" has to be both yes and no. No, God does not play dice with the world of creation with the same degree of uncertainty that we human beings experience in throwing dice. Quite irrespective of whether or not one believes in divine omniscience (God's timeless knowledge of past, present, and future), God certainly has a broader sense of what is going on in this world from moment to moment than is possible for any well-informed human being. But yes, God does roll dice in the sense of creating a world with an ever-present principle of spontaneity or creativity such as I propose in chapter two of this book.[15] God's power over creation, in other words, is more in terms of empowering creatures to make their own self-constituting decisions rather than in the sense of overpowering the creature, forcing it to conform to some preordained

---

14. See Niklas Luhmann, *Social Systems*, trans. John Bednarz Jr. with Dirk Baecker (Stanford, CA: Stanford University Press, 1996), 4–5. I analyze Luhmann's theory in chapter seven of this book.

15. See Elizabeth A. Johnson, CSJ, "Does God Play Dice? Divine Providence and Chance," *Theological Studies* 57 (1996); and my comment in the form of a *Quaesito Disputata* in the same volume of *Theological Studies* 57, 720–30: "Response to Elizabeth Johnson's 'Does God Play Dice.'" The point at issue between us was whether Transcendental Thomism or a careful blend of traditional Thomism and Whiteheadian process theology was a better vehicle to give a qualified yes to that question. As will become clear in the chapters that follow, my own position on this matter has not radically changed but only been strengthened as I see better the broad implications of a Whiteheadian process-oriented understanding of the relation between the One and the Many and the logic of universal intersubjectivity that follows naturally from it.

plan of salvation and redemption. But hopefully all this will become more evident as one reads through the chapters that follow.

Editorial Note

In elaborating his metaphysical scheme, Whitehead gave new technical meanings to many otherwise conventional words and phrases in the English language: for example, actual entity, common element of form, society. To make sure that the reader does not confuse the new meaning of the word or phrase with its more conventional meaning, I enclose each term in quotation marks at its first occurrence in each chapter. But in all further uses of the word or phrase in the chapter, quotation marks are dropped so as to facilitate easier reading of the text. For basically the same reason, I do not use quotation marks with the words Father, Son, and Holy Spirit in referring to the divine persons within the classical Christian doctrine of the Trinity. I certainly agree with Elizabeth Johnson and other Christian feminists that these words should be understood metaphorically and not literally. Feminine language could also be employed to designate the divine persons.[16] But in the interests of an easier reading of a text that is already complicated by multiple technical terms, I simply use the traditional divine names without quotation marks in the expectation that the reader will likewise appreciate their strictly metaphorical significance.

---

16. See, for example, Elizabeth A. Johnson, *She Who Is: The Mystery of God in Feminist Theological Discourse* (New York: Crossroad, 1992), 3–18.

Part One

*Philosophical Cosmology /
Natural Theology from
an Evolutionary Perspective*

1

Is There a Reason for Everything, or Do Some Things Just Happen?

In his widely read book *When Bad Things Happen to Good People*, Rabbi Harold Kushner proposes that very often bad things happen to good people for no reason at all.[1] As a result, one should not blame other people for what happened, nor blame God, nor most of all blame oneself.[2] One should instead ask God for assistance in deciding what to do next by way of making the best out of a bad situation.[3] This is excellent pastoral advice to people who are depressed or angry at what has happened to them quite unexpectedly. But it does raise questions about God and the world we live in. Kushner himself does not offer much explanation for his claim that some things happen for no reason at all beyond pointing to the fact that we live in a world in the process of evolution and that evolution is inevitably a trial-and-error process.[4] But he is also brave (or foolhardy) enough to say that there are "pockets of chaos" over which God, for the moment at least, has no control: "The world is mostly an orderly, predictable place, showing ample evidence of God's thoroughness and handiwork, but pockets of chaos remain. . . .Things happen which could just as easily have happened differently."[5] But this raises still other troubling questions. What guarantee do we have that order will ultimately triumph over chaos if God is clearly not omnipotent in dealing with creation? For that matter,

---

1. Harold Kushner, *When Bad Things Happen to Good People* (New York: Schocken Books, 2004), 53–63.
2. Ibid., 97–124.
3. Ibid., 141.
4. Ibid., 59–63.
5. Ibid., 60.

what guarantee do we have that the laws of nature will work tomorrow as well as they work today? Are we actually living on the edge of chaos all the time without realizing that this is the way things are?

Two distinguished twentieth-century philosophers of science, Charles Sanders Peirce and Alfred North Whitehead, thought seriously about such questions—namely, the relation between chaos and order, discontinuity and continuity within natural processes—and came up with remarkably similar solutions. Ironically, despite their best efforts to explain what they meant, neither one of them succeeded in persuading either the scientific community or the educated public of their day that their solutions were correct or in many cases even worthy of serious consideration. Perhaps what they proposed was too contrary to the conventional understanding of how this world works for most people to give credence to their theories. But given a fair hearing, their proposals are remarkably simple and straightforward. In brief, they both propose that the world is constituted not by people and things physically separate from one another, but by momentary energy-events with interrelated patterns of self-organization that allow these events to be linked with one another so as to constitute the persons and things of ordinary experience. From this perspective, each of us is from moment to moment a bundle of interrelated energy-events that keep our heart pumping, lungs breathing, mind working, legs and arms moving, etc. All of these events within our bodies have an instinctive feeling for one another so as to work together harmoniously. As a result, subjectivity (a feeling-level responsiveness to the environment) seems to be present not only in our human mental life but at every level of physical activity within our bodies. For that matter, it is what we share with all other creatures in this world, the feeling of acting and being acted upon at every moment of our existence.

Before launching into a brief overview of Peirce's and Whitehead's understanding of reality, however, let me return to Kushner's claim that some things just happen for no apparent reason and take note of the fact that two groups of people who basically distrust one another's intentions on virtually everything else still end up agreeing that there has to be a reason for everything that happens. In their view, nothing ever happens strictly by chance. The one group is materialists who claim that nature and nature's laws ultimately explain everything. Everything that happens is without exception based on deterministic cause-and-effect relations. The other group is religious fundamentalists who likewise claim that everything happens for a reason because God, the Cosmic Architect or Engineer, has a well-defined plan for creation. In the end, everything without exception must fit into that divine plan. So in the thinking of both groups, determinism in one form or another is the name of the game. But in my judgment, their joint claim that everything is caused by something/someone else is neither good science nor good theology, and this is why we should pay attention to what Peirce and Whitehead have to say about the nature of physical reality.

I turn now to Peirce (1839–1914), who lived and was active in philosophical circles a couple of generations earlier than Whitehead. He is customarily associated with pragmatism and later with what he himself called pragmaticism as a way to test the accuracy of scientific hypotheses. In brief, one must weigh the foreseen consequences of a concept or theory before declaring it to be even provisionally true. A concept or theory must somehow make an empirically verifiable difference if it is to be considered as worthy of acceptance and belief.[6] More important for our purposes here, however, is his view of nature or physical reality. He was strongly opposed to thinking of physical reality simply as "matter-in-motion," a cosmic machine governed by deterministic laws and principles. Instead, he believed that the laws of nature are products of an evolutionary process that is much akin to the workings of the human mind.[7] That is, just as the human mind is shaped by customary thought patterns or habits, so the laws of nature only represent statistical uniformities in the ongoing succession of physical events.[8] For example, looking at life all around us, we may say that normally B follows from A, but we cannot exclude the possibility of C, D, or E occurring instead. Hence, the laws of nature are probabilistic, not strictly deterministic in character. They predict what usually happens as opposed to what must happen in virtue of a strict cause-and-effect relation.

Studying further the workings of the human mind, Peirce also notes that it is characterized more by a flow of feeling than by a succession of thoughts. In his essay titled "The Law of Mind," he says, "Three elements go to make up an idea. The first is its intrinsic quality as a feeling. The second is the energy with which it affects other ideas. . . . The third element is the tendency of an idea to bring along other ideas with it."[9] This association of ideas on a feeling level is why habits, customary ways of thinking, develop in our minds. But, says Peirce, the same ongoing flow

---

6. Charles Sanders Peirce, *Collected Papers of Charles Sanders Peirce*, vol. 5, ed. Charles Hartshorne and Paul Weiss (Cambridge, MA: Harvard University Press, 1934), 1.

7. Charles Sanders Peirce, *Collected Papers of Charles Sanders Peirce*, vol. 6, ed. Charles Hartshorne and Paul Weiss (Cambridge, MA: Harvard University Press, 1935), 33.

8. Ibid., 97.

9. Ibid., 135.

of feeling or psycho-physical energy is at the base of habit-taking within nature. Feeling is what holds the world of people and things together at every moment.

In an essay titled "Evolutionary Love,"[10] Peirce reviews three ways to explain evolution: evolution by chance, evolution by predetermined law or mechanical necessity, and evolution through universal self-giving love. In his mind, evolution by chance and evolution by predetermined law are insufficient to explain evolution. Chance is too chaotic for any sense of long-term organization and directionality within nature to develop. Evolution by mechanical necessity does not allow for any novelty or spontaneity within nature. Only self-giving love, spontaneously giving to others as need arises rather than always protecting one's own interests, is what keeps evolution moving ahead. Females within an animal species, for example, protect their young against a predator even at risk to their own safety. Peirce here seems to raise to a cosmic principle what Jesus in the Gospels told his followers: "For whoever wishes to save his life will lose it, but whoever loses his life for my sake and that of the gospel will save it" (Mark 8:35).

What stands out here in any case is Peirce's clear linkage of what he calls the "logic of the mind" with the alleged directionality or goal orientation of the evolutionary process as a whole. Ideas, in his view, are not material realities or *things* but mental *events* rapidly succeeding one another in human consciousness. Each has its own individual moment of existence before it is succeeded by another idea, and yet there is an unmistakable feeling-level connection between them.[11] In similar fashion, says Peirce, nature as an evolutionary process exhibits three characteristics: Variety, Uniformity, and the passage of Variety into Uniformity through the formation of habits.[12] Notice how Peirce qualifies what he means by Uniformity through making reference to habits. Habits do not imply absolute uniformity or inflexible law but only regularity, a strong propensity to act in one way rather than another.

Naturally, the more spontaneity at work within an event (e.g., a moment of human consciousness versus a passing moment in the existence of a subatomic particle), the greater the power to alter old habits in favor of new ones so as to cope with changing circumstances. But the basic tendency to form habits is operative at all levels within nature. In this sense, the continuity of nature and the continuity of human consciousness are

10. Ibid., 287–317.
11. Ibid., 141.
12. Ibid., 97.

both grounded in Peirce's vision of how the world began. In the beginning there was something like chaos. But here and there patterns of organization or connections among the subatomic particles rapidly coming into and going out of existence began to take shape. With time these patterns became more complex, and, as the expression goes, the rest is history.[13]

Keeping in mind this overview of Peirce's worldview, we can now take up the philosophy of Alfred North Whitehead (1861–1947), which likewise presupposes that all of physical reality has a spiritual or psychic dimension. That is, even at the level of atoms and molecules, nature is never merely matter-in-motion but is in some sense "alive," endowed with a primitive form of subjectivity or responsiveness to the environment. As he notes at the beginning of his masterwork *Process and Reality*, "the final real things of which the world is made up"[14] are actual entities, momentary self-constituting subjects of experience. But what, you may ask, is a momentary self-constituting subject of experience? Whitehead replies: Look into your ongoing consciousness and see what is happening at every waking moment. Our senses are giving us empirical data from our bodies and through our bodies from the outside world. Somehow we always find a way subconsciously to organize that multiplicity of data within our consciousness so that we end up with ourselves as the subjective observer of the objective world around us. Whitehead then claims that this unconscious process of bringing data together into a unified whole is also taking place everywhere else in nature.

That is, all animals and plants, even all inanimate things, are made up of what scientists call subatomic particles. These particles, however, are not tiny bits of matter that have existed from the very beginning of the universe as a result of the big bang. No, according to Whitehead, a subatomic particle is from moment to moment a new self-constituting subject of experience, feeling the world around it and in some fashion responding to it. So what to human observers appears to be a material entity or "thing" is actually a "society," an ongoing series of such momentary subjects of experience with basically the same pattern of internal organization and external activity.[15] These spiritual atoms are then psycho-physical events with an "inside" as a momentary self-constituting subject of experience and an "outside" in terms of a stable minipattern of organization and

---

13. Ibid., 33.
14. Alfred North Whitehead, *Process and Reality: An Essay in Cosmology*, corrected edition, ed. David Ray Griffin and Donald W. Sherburne (New York: Free Press, 1978), 18.
15. Ibid., 34.

activity. Furthermore, just as material atoms are commonly thought to aggregate into molecules with some molecules becoming components in living cells, so these spiritual atoms, according to Whitehead, aggregate into "societies," collections of momentary subjects of experience with a feeling for one another and thus with a built-in tendency to form bonds with one another. Such societies at an advanced stage of complexity then become the persons and things of common sense experience.

But how can something immaterial, a subject of experience, even in combination with other such immaterial subjects of experience, become something tangible, a person or thing that I can see, hear, touch, etc.? Whitehead's answer is thought provoking. When we look around the world, we admittedly see only things, some living and many more non-living; we do not see subjects in dynamic interaction. Whitehead argues, however, that our ordinary perception of things is deceptive because it is based on an unconscious process of abstraction from, and consequent simplification of, an enormous amount of feeling-level data available to us at every moment in virtue of our senses and central nervous system. That is, we first experience a multitude of sense impressions (e.g., colors, sounds, smells) that convey to us the presence and activity of other subjects of experience in our vicinity. This is what Whitehead calls "causal efficacy."[16] A split second afterward, in virtue of what he calls "presentational immediacy," we subconsciously represent to ourselves in imagination the mental world described above, namely, full-color images of all the things and persons from whom these feelings came in the first place.[17] So strictly feeling-level energy-events through the unconscious workings of the human mind become the people and things of conventional sense perception.

Whitehead, accordingly, is reminding us how complex the world of common sense experience is just under the surface. At every moment between the rest of the world and ourselves there exists a complicated network of psycho-physical events that link us to the rest of the world and the rest of the world to us. We think, for example, that we are the same person as yesterday, but the events going on inside us and around us in the intervening time have made us a somewhat different person today: certainly a day older, perhaps a day wiser, but in any case subtly different than we were yesterday. In the same way, the world as a whole is an ongoing sequence of psycho-physical events with basically the same pattern of organization and activity as the moment before but still in some modest

---

16. Ibid., 116, 169.
17. Ibid., 169.

way different. This is, after all, what it means to live in a world constituted by ever-changing events rather than unchanging things.

Notice how Whitehead's theory gives greater specification to Peirce's proposal that the "law of the mind" is at the same time a law of nature. For if everything that exists is made up of momentary subjects of experience in dynamic interrelation, then it is obvious that nature as a whole obeys what Peirce calls the law of mind. Hence, not just the human mind but nature as a whole is based on habits, customary patterns of behavior, rather than on rigidly deterministic laws. That is, these habits prescribe what regularly happens rather than what must happen at any given moment so as to allow the entity in question at least at intervals to undergo a change of pattern or mode of operation in line with changes in the environment.

Likewise, Peirce's claim that evolutionary love rather than chance or mechanical necessity is the driving force of evolution makes a lot of sense within Whitehead's metaphysical scheme. Material things, after all, do not feel one another's presence and activity, but subjects in dynamic interrelation do respond to one another on a feeling level. So internal feelings rather than impersonal external forces or pure chance keep the process of evolution going. Likewise, granted the reality of evolutionary love as the driving force in evolution, there is no need to appeal to God for periodic active intervention to keep the world from collapsing into chaos. Rather, moved by evolutionary love and under subtle divine inspiration, as I will explain shortly, nature can be trusted to be self-organizing and self-correcting.

Beforehand, however, I want to offer one modest correction to Whitehead's scheme, which in my judgment aligns it even more closely to Peirce's understanding of reality. In my view, Whitehead's notion of societies as aggregates of actual entities or momentary subjects of experience with basically the same pattern of self-organization is a little too simplistic. A Whiteheadian society, after all, should have an objective unity over and above its component parts, namely, individual self-constituting subjects of experience in their dynamic interrelation. The whole, in other words, should be not only *quantitatively more* but also *qualitatively other* than its component parts. My argument for many years now has been that Whiteheadian societies should be understood as fields of activity or environments structured by the interplay of ever-new generations of these momentary subjects of experience.[18] In this way, subjects of experience

---

18. See Joseph A. Bracken, *The One in the Many: A Contemporary Reconstruction of the God-World Relationship* (Grand Rapids, MI: Eerdmans, 2001), 131–51; *Subjectivity, Objectivity, and Intersubjectivity: A New Paradigm for Religion and Science* (West Conshohocken, PA: Templeton Foundation Press, 2009), 124–37.

do not originate within a vacuum but arise within an already structured, energy-filled environment. Furthermore, upon passing out of existence in the next instant, they contribute the energy-content and structure of their own individual self-constitution to the enduring energy-content and structure of the surrounding field of activity or environment. In this way, the continuity of existence for things so much emphasized by Peirce is likewise assured for a Whiteheadian society since it endures with virtually the same basic pattern of organization even as its constituent subjects of experience arise and perish at every moment.

So what have we accomplished with this moving back and forth between the worldviews of two prominent philosophers of science in the early twentieth century? In a nutshell, combining the schemes of Peirce and Whitehead in this way allows me to propose, in line with St. Paul's speech to the Athenians in the Acts of the Apostles 17:28, that we and all of God's creatures "live and move and have our being" within God. We share with the three divine persons of the Christian doctrine of the Trinity a common living space within which to work out our relationships both to one another and to the divine persons. In biblical terms this common space can be called the kingdom of God.[19] But it is not for that reason a fixed and unchanging reality. Rather, it is something dynamic and ever changing since our relationships to one another and to the divine persons are based on habits capable of modification over time, not on deterministic laws of nature or on some predetermined divine plan for our salvation.

But just how do we live, move, and have our being within God so that we share a "space" with the three divine persons as coparticipants in the kingdom of God? First of all, as already indicated, we have to agree with Whitehead that the world is made up not of physical atoms but of spiritual atoms, momentary self-constituting subjects of experience in dynamic interrelation. So the world we live in is radically intersubjective; in the end, there exist only societies or complex groupings of interdependent subjects of experience. Then, if one accepts my modification of Whitehead's scheme to the effect that these societies are not simply aggregates of subjects of experience but rather invisible fields of activity or environments that are structurally the same from moment to moment, it is easy to make a series of claims about the God-world relationship from a Christian perspective.

Our first claim is that the three divine persons of the Christian doctrine of the Trinity as subjects of experience in ongoing dynamic interrelation coconstitute a single infinitely extended field of activity proper to them-

---

19. Joseph A. Bracken, *Christianity and Process Thought: Spirituality for a Changing World* (Philadelphia, PA: Templeton Foundation Press, 2006), 53–64.

selves as one God. Secondly, this divine field of activity is then the ground or source of the world of creation, given the free decision of the divine persons to share their divine life with creatures.[20] The big bang, in other words, with which creation began fourteen billion years ago, presumably took place *within* God, that is, within the all-encompassing divine field of activity coconstituted from all eternity by the ongoing activity of the divine persons vis-à-vis one another. This is my own understanding of the traditional Christian doctrine of creation out of nothing (*creatio ex nihilo*). The divine field of activity is "no-thing," that is, not a thing or an entity but a context or an environment for the "things," the divine persons and all their creatures as ongoing interrelated subjects of experience, existing within it.

Thus, at the beginning of creation, much as Peirce claimed about the relatively chaotic origin of the universe, there were only momentary subjects of experience coming into and going out of existence very rapidly and in total isolation. But eventually the habit-forming tendency took over and societies or structured fields of activity for these momentary subjects of experience took shape, with the result that initially subatomic particles, then atoms, and finally molecules were formed. Slowly but surely, then, the world as we know it today began to emerge as a reality distinct from God but still within the all-encompassing divine field of activity and thus always under the inspiration or prompting of the three divine persons in terms of what Whitehead calls divine "initial aims," feeling-level lures for decision or action on the part of the creature.[21]

Skipping over the details of our fourteen-billion-year cosmic history so as to focus on ourselves here and now as the current inhabitants of planet Earth, we can say that we live in a world constituted not by things physically separate from one another but by multiple partially overlapping and hierarchically ordered fields of activity. Each of us as an individual, for example, is what Whitehead calls a "structured society," a society made up of subsocieties of actual entities or momentary subjects of experience arranged in a hierarchical order.[22] The uppermost society or field of activity for each of us is our soul or, in its conscious moments, our mind, will, and feelings. Supporting our soul is our body, still another very complex structured society beginning with the brain as the endpoint and focus of our central nervous system, which links the brain to all the other bodily organs or subfields of activity within the body. Each of these subfields has

20. Ibid., 3–13.
21. Whitehead, *Process and Reality*, 244.
22. Ibid., 99–100.

its own integrity within the body, even as each contributes to the overall functioning of the soul-body unity, which is you or me as a living and breathing human being.

But as human beings, we are likewise participants in many socially organized fields of activity around us: the family, the local community with its multiple forms of political and economic organization, the church and the nation as still more complex fields of activity to which we belong and in which we participate, and finally, the rest of the world with which we are in regular contact through radio, television, the internet, etc. Each of these overarching socially organized fields of activity has its impact on us and our lives as individuals here and now, and we through our decisions from moment to moment have some modest effect on the structure and operation of these broader fields of activity. Needless to say, such a scheme for understanding our place in the world around us also has tremendous implications for ecotheology, the new understanding of the God-world relationship with conscious focus on saving the planet for future generations to enjoy.

Finally, to return to our original claim that the world of creation came into existence and still continues to exist within the all-inclusive field of activity proper to the divine persons in their own divine life, we can say that even now we share in their divine life and they share in our lives without our normally realizing it. For what we say and do has an impact on the divine persons both in their relations to one another and in their relations to us as their creatures. Similarly, they never cease exercising care or providence over us and the world in which we live. They do so, however, not as in classical theology by setting up a plan of action that we and all other creatures are expected to follow under pain of punishment, either now or hereafter. Rather, given that we too are independent subjects of experience with a built-in power of self-determination at every moment, the divine persons use initial aims, feeling-level lures to action in one direction rather than in another. We are free, of course, to reject or at least to significantly modify these feeling-level urges and desires coming from the divine persons and in that way to introduce disorder into our own personal world and into the broader world of the various communities or organizations to which we belong.[23] But as Jesus made clear in his life and message and above all in his passion, death, and resurrection, the divine persons never seem to give up on us even in our most foolish moments; they always have a new initial aim, a new "actual grace," ready for us to

---

23. Bracken, *Christianity and Process Thought*, 77–88.

help us deal with any unhappy situation that we may have created by our ill-fated decisions of the moment.

Transposed to a cosmic perspective, this means that evolution, even as the strictly trial-and-error process that it has turned out to be, is the divinely chosen way first to create and then to sustain this world of ours. Perhaps this is why it has taken us fourteen billion years to get to where we are now; who knows how long it will take for this world of ours to achieve whatever the divine persons originally had in mind for it at the start of the cosmic process? It is our Christian hope, of course, that someday we will see with the divine persons how it all worked out, for ourselves personally, for all other human beings, for the earth on which we live, and for the entire universe as a vast and incredibly complicated network of interdependent processes. For now, however, we may join the medieval mystic Julian of Norwich in repeating to ourselves over and over again: "All will be well; all will be well."[24] Human freedom and divine providence are not incompatible. Unexpected events do happen in this world of ours simply by chance, but in the end God's will and whatever we creatures spontaneously choose to do will be reconciled, and all will be well.

---

24. Julian of Norwich, *Showings*, ed. Edmund Colledge, OSA, and James Walsh, SJ (New York: Paulist Press, 1978), 153.

2

Value and Creativity

Before one makes value judgments about specific lines of research in science and particular applications of scientific research to technology, one should have some preunderstanding of what is meant generically by the term "value." For example, is value ultimately based on broad consensus with respect to subjective desires and purposes or on something objective in the natural order to which appeal can be made in evaluating the merit of various value judgments? In this chapter, I will briefly sketch an ontology of value based on an evolutionary worldview as opposed to a more classical understanding of value arising out of a God-given or some other relatively fixed plan or design for the way that the world should operate. My thesis is that creativity is the origin of value and that the degree of creativity at work in individual cases is the basis for objective value judgments.

This has some affinity with Stuart Kauffman's proposal in his recent book *Reinventing the Sacred* that creativity is the most appropriate symbol for reinventing (or, I would say, rediscovering) a sense of the sacred in modern life.[1] But whereas Kauffman seems to regard creativity as Ultimate Reality in its own right, I would side with Alfred North Whitehead in his book *Science and the Modern World* that God is "the principle of limitation" for the operation of creativity, given that creativity is so unpredictable in its ceaseless activity.[2] Whitehead would agree with Kauffman that creativity is "the universal of universals characterizing ultimate matter

---

1. Stuart Kauffman, *Reinventing the Sacred: A New View of Science, Reason, and Religion* (New York: Basic Books, 2008), 281–88.
2. Alfred North Whitehead, *Science and the Modern World*, 2nd ed. (New York: Macmillan, 1967), 179.

of fact."³ But more precisely than Kauffman, Whitehead likewise makes clear that creativity is not in any sense an entity (e.g., nature as a whole for Kauffman⁴) but simply a principle of activity that is actual only in the entities that it enables to exist.⁵ It is, in other words, the hidden nature of things, their Aristotelian *physis*, which empowers them to be creative both in their own self-constitution and in their impact upon other entities. In this respect, creativity is the equivalent in an evolutionary worldview of the concept of being in classical metaphysics, provided that one understands being as a verb or participle rather than as a noun. In both cases, a principle of activity is what is primarily intended. Thomas Aquinas, to be sure, seems to have understood being primarily as a noun rather than as a verb or participle. For example, the philosophical description of God for Aquinas is *ipsum esse subsistens*.⁶ Grammatically, *esse subsistens* (subsistent being) is a noun, not a verb. In my view, this is an unintentional confusion of what exists (entity) and that by which something exists (its nature or essence). However, since God for theists is believed to be transcendent of being in the conventional sense and yet the origin or source of all created beings, one does not readily notice an otherwise obvious distinction.

Ironically, this logical mistake allowed medieval theologians like Aquinas, influenced no doubt by the writings of Plotinus and other neoplatonists, to establish a graded hierarchy of entities, from God as pure actuality to prime matter as pure potentiality. This, in turn, allowed them to set up a relatively fixed hierarchy of values. That which has more being has more value. Value is based on actuality rather than potentiality. Potentiality is equivalently a disvalue until it achieves actuality in terms of the perfection of its nature or essence. Change thus constantly takes place in the natural world but only within predetermined limits. Individual entities come into existence, endure for a while, and then pass out of existence; but the basic structure of the world remains unchanged with the passage of time. Moreover, relationships between entities are contingent events, "accidents" with respect to the enduring substantial reality of those same entities.⁷ Only with respect to the doctrine of the Trinity in his *Summa*

---

3. Alfred North Whitehead, *Process and Reality: An Essay in Cosmology*, corrected edition, ed. David Ray Griffin and Donald W. Sherburne (New York: Free Press, 1978), 21. Cf. Kauffman, *Reinventing the Sacred*, 287–88.

4. Kauffman, *Reinventing the Sacred*, 288.

5. Whitehead, *Process and Reality*, 31.

6. Thomas Aquinas, *Summa Theologiae* (Madrid: Biblioteca de Autores Cristianos, 1955), I, q. 7, a. 1.

7. Aristotle, *Metaphysics*, in *The Basic Works of Aristotle*, ed. Richard McKeon (New York: Random House, 1941), 1088a.

Theologiae did Aquinas concede that relationships between entities (the divine persons) are constitutive of their very being or existence.[8] But the doctrine of the Trinity in the *Summa* was never integrated with his overall understanding of the God-world relationship, which remained governed by the Aristotelian categories of substance and accident.

When, however, being is conceived as a verb or participle and thus as a principle of activity equivalent to the modern-day understanding of creativity, then one's worldview is dramatically altered. Entities are graded and valued according to their potentiality, not their current actuality. God is no longer conceived simply as the Supreme Being or Pure Unchanging Actuality but also as the entity with the greatest potentiality for further actualization. Thereby I do not claim here that God is Pure Becoming or simply a principle of activity with no entitative status in its own right (akin to Kauffman's understanding of creativity). As Whitehead comments in his masterwork *Process and Reality*, Creativity as an ontological principle exists only in its instantiations, actual entities of one kind or another.[9] So God must be an entity, yet an entity possessing much more creativity than any other entity. Furthermore, if, as I believe, God (*ipsum esse subsistens*) possesses Creativity by nature or in its fullness and is not a "creature" of Creativity as Whitehead himself maintains,[10] then it follows that finite entities possess a measure of creativity by participation in the divine nature or act of being. In proportion as finite entities possess varying degrees of creativity, they can be aligned into a new hierarchy of being, this time understood as a hierarchy based on potentiality rather than actuality. That is, beings have value in proportion to their inbuilt potentiality or capacity to adapt to their environment and further evolve. What is important and valuable, therefore, is not what an entity is at the present moment but what it can in due time become in virtue of the potentiality or creativity at its disposal. Yet at all levels of the hierarchy, creativity works in the same way; that is, it enables a given entity in its current state to deal successfully with its environment and thereby in some modest way to change or evolve internally with respect to its nature or essence. One might see in this understanding of creativity a philosophical explanation of the Darwinian principle of natural selection in the realm of biology. But in my judgment it also offers a philosophical explanation of the inbuilt self-organization of entities at all levels of existence and activity within Nature (as I shall explain more fully in chapter three).

8. Ibid.
9. Whitehead, *Process and Reality*, 7.
10. Ibid., 88.

Yet precisely in this widespread use of creativity as a principle of change or evolution, still another qualification must be made, a reservation that Kauffman in his appreciation of creativity seems to have downplayed. As noted above, Whitehead claims in *Science and the Modern World* that creativity is morally neutral since in principle its effects can be destructive rather than productive, at least in the short run.[11] Only God at work in this world through the principle of creativity can be counted on to use its power consistently for productive, not destructive, purposes. All the finite entities of this world are capable of a destructive use of their inbuilt creativity, if only because they do not see the full consequences of their self-constituting "decisions" at any given moment. They are blind (or in some cases they blind themselves) to the harm that they can cause both to themselves and to other entities through an inappropriate use of the creativity entrusted to them by God. Here, to be sure, I am presupposing still another principle of Whitehead's metaphysical scheme, namely, that "the final real things of which the world is made up are actual entities,"[12] momentary subjects of experience that make themselves to be what they are here and now by a self-constituting decision. This is not to claim, of course, that such decisions are consciously made. Without a brain and a corresponding central nervous system, an entity cannot be conscious of itself and its environment. So atoms, molecules, and lower forms of life interact with their environment in a highly predictable, more or less mechanical way. Only members of higher-order animal species and, above all, human beings can make conscious decisions in virtue of their innate power of creativity. Yet here too, Whitehead would claim, the degree of consciousness in human—and even more so in nonhuman—decision making is quite limited.[13] That is, even human beings make most decisions intuitively or semiconsciously simply as a spontaneous response to some event happening around them in the external environment and subsequently inside themselves by way of emotional reaction to that event which commands their attention.

Even when creativity is used for destructive purposes, however, it still has some limited value as an exercise of creativity, the actualization of an unrealized or previously unknown potentiality. Moreover, the assumption here has to be that in the long run something currently quite destructive can, if properly dealt with, have an overall constructive effect on the cosmic process as a whole. This is, after all, still another way in which a value system based on the principle of creativity or potentiality is quite

---

11. Whitehead, *Science and the Modern World*, 179.
12. Whitehead, *Process and Reality*, 18.
13. Ibid., 106–9.

different from classical value systems based on a principle of actuality. The value of creativity, in other words, is grounded more in aesthetics than in ethics. The ultimate criterion for what is valuable is not what is ethically good or bad as an actuality here and now. Rather, the enduring value of an event is to be judged in terms of the way that it actualizes a hitherto untried potentiality and in some measure contributes to a still unfinished cosmic process. Thus even what is prima facie destructive in terms of its consequences (e.g., a hurricane or earthquake in a heavily populated area of the world) can be valued as a powerful manifestation of the creativity inherent in the cosmic process. At the same time, for human beings there also exists a heavy moral responsibility for dealing effectively with the consequences of such a catastrophic event both for themselves and for other human beings. So while the good is ontologically subordinate to the beautiful within this scheme, the need for moral goodness, especially in times of crisis, is not thereby diminished. This aesthetic approach to the problem of evil perhaps makes clear what classical Taoism has maintained for centuries. Judgments of good and evil on the part of human beings are important for the situation at hand, but in terms of the cosmic process as a whole they are inevitably perspectival.[14] Unlike a Creator God, human beings can never make value judgments based on the overall workings of the cosmic process but only on their limited understanding of the consequences of that process for themselves and other human beings here and now. Action to deal with the situation is clearly needed, but this action too will have consequences, both good and bad, that cannot be fully anticipated by the individuals here and now directly involved.

A few years ago, Christopher Southgate published a thought-provoking book on the inevitable negative consequences of the Darwinian principle of natural selection in a predator-prey natural world.[15] His focus was on nonhuman or animal suffering rather than on human suffering as a result of cosmic evolution because most Christian philosophers and theologians deal with the problem of suffering from a strictly human point of view.[16] In this way they overlook all the suffering involved in the progressive extinction of animal species within the cosmic process so as to allow for the growth in bodily and mental complexity needed for the eventual emergence of our own human species. Southgate, however, takes very seriously this

---

14. *Religions of the World*, 3rd ed., ed. Robert K. C. Forman et al. (New York: St. Martin's Press, 1993), 236.

15. Christopher Southgate, *The Groaning of Creation: God, Evolution, and the Problem of Evil* (Louisville, KY: Westminster/John Knox, 2008).

16. Ibid., x.

challenge to traditional Christian belief in the goodness of God when the price for the emergence of the human species in the evolutionary process is so high in terms of innocent suffering on the part of nonhuman species for the achievement of a value totally divorced from their own well-being here and now.[17] His response in terms of a firm belief in a trinitarian Creator God is understandably quite nuanced. But briefly stated, he claims:

> Out of the "futility" of the evolutionary process, and the extinction of over 98 percent of the species that have ever lived, came, precariously and eventually, "hope." The rhythm of nature's birthing and dying, with all the creaturely suffering that we have seen necessarily attends it, awaited the ultimate self-transcendence of the humanity of the Christ, whose dying and rising again inaugurated a new era of possibilities.[18]

What Southgate means by "a new era of possibilities" is not only the redemption of the human race in terms of life after death but also the survival in some form or other of nonhuman animal species in the *eschaton*, the fullness of God's kingdom at the end of the cosmic process.[19] As I see it, Southgate's sensitive study of the evil caused to nonhuman animal species as a consequence of the evolutionary process confirms my contention that the problem of evil within the divine plan of Creation and Redemption must ultimately be judged more in terms of aesthetics than of ethics. That is, the value of the self-giving love of the triune God in the person of Jesus and the often halting response of creatures to that self-giving love ("the groaning of creation") cannot be assessed in strictly rational terms as a decision for good over evil but only on an aesthetic basis as a totally unexpected expression of divine and creaturely self-transcendence.

In addition, from a practical point of view, this long-range aesthetic approach to value judgments might be useful in resolving issues related to theodicy. Literally, theodicy means justification of God's ways of dealing with the world and, above all, with human beings in their periodic trials and sufferings. In point of fact, however, theodicy ends up being a justification of our human understanding of God's wisdom and goodness with respect to creatures. As Harold Kushner points out in his classic work *When Bad Things Happen to Good People*, we human beings feel an urgent need to give a reason or find an underlying cause for everything that happens to us and around us. If something goes wrong, someone has to be

---

17. Ibid., 13.
18. Ibid., 94–95.
19. Ibid., 89.

blamed.[20] Our sense of good and bad, however, is inevitably perspectival, that is, related to ourselves or to the group to which we belong here and now. A willingness to allow the unexpected to happen and to appreciate the spontaneity that brought it into being, even when its initial consequences present a serious challenge to the status quo, is an attitude hard to sustain when our instinctive response to the event is strongly negative. But this may be the mode of operation of a Creator God who sees the long-term workings of the cosmic process as well as its short-term consequences for particular individuals. God is more patient than we human beings are in dealing with natural selection or its equivalent not only in biology but also in the world of nature at large. Unexpected or spontaneous change is sometimes a pleasant experience, but more often it is painful to those directly involved. Yet without the ever-present possibility of this kind of change, the status quo would never vary, and the cosmic process would be at a dead end.

Human value judgments, therefore, should be assessed as far as possible in terms of the long-term consequences of a present decision. But how does one calculate these long-term consequences? Charles Sanders Peirce believed that the answer to this question was to be found in his notion of patterns or habits (more exactly, the phenomenon of habit making). Patterns seem to be at work everywhere in the natural world as well as in the minds of human beings. Matter, as he sees it, is "effete mind"[21]—that is, mind utterly lacking in spontaneity—with "inveterate habits becoming physical laws."[22] As we have already seen in the last chapter, Alfred North Whitehead offered an explanation along the same lines as Peirce with his notion of a "society" as an assemblage of actual entities (momentary self-constituting subjects of experience existing in space and time) with a "common element of form,"[23] that is, a more or less uniform pattern of relation to one another. So out of the apparent randomness of the contingent decisions of these "spiritual atoms" at any given moment, a new social order or a bigger corporate reality inevitably emerges. New patterns of activity and behavior among these spiritual atoms produce in the end the persons and things of common sense experience. More on

---

20. Harold Kushner, *When Bad Things Happen to Good People* (New York: Random House, 2004), 9–35.

21. Charles Sanders Peirce, *Collected Papers of Charles Sanders Peirce*, vol. 6, ed. Charles Hartshorne and Paul Weiss (Cambridge, MA: Harvard University Press, 1935), 158.

22. Ibid., 25.

23. Whitehead, *Process and Reality*, 34.

Whitehead's metaphysical scheme will be forthcoming in due time. For now, it is only important to acknowledge that present decisions by actual entities (momentary subjects of experience) shape both the past and the future of the societies to which these actual entities belong.

For example, when a young man and a young woman decide to get married, that decision reorders the past history of each of them as an individual; that is, they will each look upon their past lives as somehow leading up to their getting acquainted, falling in love, and deciding to get married. Similarly, from this moment on, they will be facing the future as a married couple making joint decisions. So the future in terms of possibilities is radically altered for both of them in virtue of their current decision.[24] Not every human decision, of course, has such a decisive impact on the individuals who here and now make it. Furthermore, human beings have far more freedom of choice than other higher-order animals with respect to past decisions and in view of future possibilities. They can set up a new pattern of thinking and behavior that alters or abandons altogether an older pattern. Nevertheless, for all sentient beings as well as for human beings, the experience of time, however rudimentary, is an organic unity; past, present, and future are dynamically interrelated. Likewise, at all levels of existence and activity within Nature, value is grounded in individual subjectivity, that is, in the momentary self-constituting decision of an actual entity; but individual subjectivity only attains objectivity through intersubjectivity, namely, through repetition of that same pattern of self-constitution in subsequent actual entities belonging to the same society over space and time.[25] Thus, value in the created order is never imposed on an individual subject of experience by an outside source or influence but is consciously or unconsciously generated by its decision here and now, which then one way or another is repeated by its successors in the same society. This is not to deny the possibility of transcendent values or ideals at work within the cosmic process (see my remarks on divine "initial aims" below), above all in the minds and hearts of human beings. But at

---

24. See Joseph A. Bracken, *Christianity and Process Thought: Spirituality for a Changing World* (Philadelphia, PA: Templeton Press, 2006), 79–80; see also Robert Cummings Neville, *Eternity and Time's Flow* (Albany, NY: State University of New York Press, 1993), 111.

25. See Joseph A. Bracken, *Subjectivity, Objectivity, and Intersubjectivity: A New Paradigm for Religion and Science* (West Conshohocken, PA: Templeton Foundation Press, 2008). This is my basic response to the problem of subjectivity and objectivity in the history of Western philosophy. It follows necessarily from my new paradigm for the relation between the One and the Many.

any given moment, the value thus achieved remains finite and therefore subject to further evolution in terms of its scope and significance.

Having indicated in a general way what I see as the linkage of value and creativity, in what follows I sketch a preliminary worldview in which creativity as a value-creating ontological principle plays a key role. It will be largely based on the cosmology of Alfred North Whitehead, which we have already seen in chapter one. But as already noted there, it differs from Whitehead's own understanding of society as an assemblage of actual entities, momentary self-constituting subjects of experience, with a common element of form.[26] For Whitehead failed to indicate how a society with its common element of form will normally endure even while its component actual entities succeed one another with such great rapidity. For what is to guarantee that a society will not dramatically change form as a result of new constituents in the next moment of its existence and thus become a new kind of entity? My claim for many years now has been that a Whiteheadian society is better understood as an environment or structured field of activity for its constituent actual entities. This environment or structured field of activity does indeed evolve or change character with the passage of time, but from moment to moment it offers an enduring law-like context for the emergence and self-constitution of successive sets of actual entities. Somewhat akin to the Aristotelian notion of "substance,"[27] a Whiteheadian society is a necessary principle of continuity in a world marked by ongoing change. But unlike an Aristotelian substance, the society also changes in its basic structure or form in and through the interplay of its constituent actual entities, albeit at a much slower rate than those same constituent actual entities, which come into and go out of existence so quickly. In effect, then, a society of actual entities and its constituent actual entities necessarily condition one another's existence. A society would be an empty field of activity and thus meaningless without the rapid succession of its constituent actual entities. Yet the actual entities at any given moment are heavily conditioned in their self-constitution by the already existing law-like context of the environment in which they arise. Much like the relation between form and emptiness in the Buddhist tradition,[28] therefore, actual entities and the societies to which they belong are intrinsically interdependent realities.

26. Whitehead, *Process and Reality*, 34.
27. Aristotle, *Metaphysics*, 1028a.
28. Donald W. Mitchell, *Buddhism: Introducing the Buddhist Experience* (New York: Oxford University Press, 2002), 101.

Within this value-oriented worldview, the entity with the greatest potentiality is also the Supreme Actuality, namely, God. God, after all, has the greatest capacity to both affect and be affected by everything else that exists. In this way, the potentiality within God is unsurpassed, even though from moment to moment God is likewise the Supreme Actuality for that moment. Furthermore, the unity of God is not undifferentiated or "simple," as in the philosophy of Thomas Aquinas;[29] rather, in keeping with the nature of God as creativity, the inner unity of God is dynamically differentiated. That is, in line with classical Christian doctrine, God is triune, a dynamic unity of three closely interrelated divine persons who are together one God. In Whiteheadian terms, God as thus understood is a "structured society," a society composed of subsocieties of actual entities.[30] Each of the divine persons is what Whitehead calls a "personally ordered" society of divine actual entities (momentary subjects of experience) with an ongoing distinctive pattern of existence and activity. Hence, their unity as one God is a structured society with its own distinctive common element of form or pattern of existence and activity as a higher-order corporate reality. Christian tradition has always maintained that each of the divine persons is God but in a totally different way. To be God as Father is completely different from being God as Son or as Holy Spirit. Only the dynamic fusion of these radically different ways of being the one God constitutes what is meant by "God" in the Christian tradition.

Furthermore, if one accepts my modification of Whiteheadian societies as structured fields of activity for their constituent actual entities, then each of the divine persons has its own field of activity which in principle is infinite or unlimited in scope. But if all three of these interrelated fields of activity are thus unlimited or infinite, then their unity as a structured society can only be a single field of activity common to all three persons. Elsewhere I have referred to this all-embracing divine field of activity as the divine matrix or the ontological ground of existence and activity for all other entities in this world.[31] Within this divine matrix our universe presumably came into existence roughly fourteen billion years ago. Through a free decision of the divine persons, there was a transfer of divine energy or creativity to a new creaturely actuality within the all-encompassing divine field of activity. What happened next was what natural scientists conventionally call the big bang, an enormous explosion of energy in all

---

29. Aquinas, *Summa Theologiae*, I, q. 3, a. 7.

30. Whitehead, *Process and Reality*, 99.

31. Joseph A. Bracken, *The Divine Matrix: Creativity as the Link between East and West* (Maryknoll, NY: Orbis Books, 1995).

directions. Yet since creativity is the functional principle whereby "the many become one and are increased by one,"[32] after the initial explosion that produced an innumerable number of energy-events or "virtual particles," a gradual process of unification of these momentary "particles" or transient energy-events took place first into primitive enduring societies (protons, neutrons, electrons) and then by degrees into even more complex social units (atoms and eventually molecules of varying size and complexity). Thus, divine creativity as shared with finite subjects of experience gradually brought into existence higher-order societies corresponding to the inanimate and animate "things" of this world. Furthermore, these higher-order groupings of actual entities, I would argue, emerged through a combination of what might be called bottom-up and top-down causality.[33] That is, the constituent actual entities in each of the lower-order societies by their dynamic interrelation from moment to moment represent bottom-up efficient causality in the emergence of a higher-order society. But the resultant common element of form for that higher-order society then exercises top-down or "informational" causality upon all subsequent sets of actual entities in the society so that it can continue to exist with basically the same element of form or governing structure.[34]

In its constitution, therefore, a Whiteheadian society exemplifies what might be called "emergentist monism."[35] All its constituent parts or members are made of the same "stuff." But the society is thus not simply reducible to the ongoing interaction of its constituent parts or members. It is a further physical reality that is necessarily coexistent with its constituent parts or members, and yet it is both more than and other than those same constituents. More specifically, it is the necessary environment or already structured field of activity for these actual entities in their emergence and dynamic interrelation at any given moment. Understood in this way, a Whiteheadian society is a clear instance of strong versus weak nonreductive physicalism.[36] That is, as indicated above, within a society there is genuine top-down, as well as bottom-up, causality. As Whitehead makes clear in *Process and Reality* with reference to the mind-body relation within human beings, the human brain at every moment receives the information coming

---

32. Whitehead, *Process and Reality*, 21.
33. Bracken, *Subjectivity, Objectivity, and Intersubjectivity*, 138–53.
34. Ibid., 159–63. See also John Polkinghorne, *The God of Hope and the End of the World* (New Haven, CT: Yale University Press, 2003), 8–10.
35. Philip Clayton, *Mind and Emergence: From Quantum to Consciousness* (New York: Oxford University Press, 2004), 60–62.
36. Ibid., 1–37.

from the body as a coordinated set of muscles, veins, and vital organs; then it responds in terms of its own mental history (memories of previous experiences with accompanying feelings of either fear or attraction). The resulting synthesis of feelings coming from the mind and body result in a conscious, or more often unconscious, decision by a human being that impacts the body and brain at the same time.[37]

After reflecting upon this interaction of mind and body within a human being from moment to moment, one might possibly conclude that Whitehead implicitly endorses a dualism of soul and body, spirit and matter. This, however, is not the case, since every actual entity is in his mind both an immaterial and a material reality at the same time, what he calls a "subject-superject."[38] As a momentary self-constituting subject of experience, it is a strictly spiritual reality; but as a superject, it is a physical reality, a structure or pattern that expresses the self-constituting decision of an actual entity. This structure or pattern can then be "prehended," actively felt or perceived, in subsequent moments of experience.[39] Thus, neither spirit nor matter can exist without the other. Together they constitute the full reality of every entity in this world (from subatomic particles to the human mind and the social institutions that human beings have created over time). At lower levels of existence and activity within Nature, the constraints of matter on the activity of spirit are considerable, so that, as Peirce claims, matter is "effete mind," utterly lacking in spontaneity.[40] But at higher levels of existence and activity (e.g., human beings and the corporate institutions they create), the flexibility and spontaneity of spirit come more and more to the fore.[41] One might object, of course, that this position still represents a form of panpsychism, the belief that "it's mind all the way down," empirical evidence for which is lacking or at least quite ambiguous.[42] But this objection is based on the assumption that between monism and dualism there is no middle term or common ground. The doctrine of nondualism (which is widely accepted in classical East Asian philosophies like Advaita Vedanta Hinduism, Buddhism, and Taoism) counterproposes that between contraries, as opposed to contradictories, there can and should exist a middle-ground position that mediates between them.[43]

---

37. Whitehead, *Process and Reality*, 109.
38. Ibid., 45.
39. Ibid., 28.
40. Peirce, *Collected Papers*, vol. 6, 158.
41. Bracken, *Subjectivity, Objectivity, and Intersubjectivity*, 160–61.
42. Clayton, *Mind and Emergence*, 130.
43. Cf. David Loy, *Nonduality: A Study in Comparative Philosophy* (New Haven, CT: Yale University Press, 1988), 178–86.

In any event, because a Whiteheadian society has its own ontological identity in distinction from the self-constitution of its individual constituent actual entities at any given moment, it has a value that is not only more than but likewise other than the fleeting value of an individual subject of experience in its momentary self-constitution. A society endures over time and thus has value precisely as something that endures and is not irrevocably lost in the next moment of the cosmic process. At the same time, its ontological value is basically self-generated, not imposed from the outside by some external standard of value emanating from God as the transcendent source of all value or from the contingent value judgments of fallible human beings by way of consensus. A society has an objective value simply as a consequence of what it has become in and through its succession of constituent actual entities. Naturally, some societies have more value than others in terms of their structure and complexity as well as in view of the role that they play in the constitution of still other much larger societies. A unicellular organism like a virus, for example, has less value than the multicellular organism (e.g., a human body) that it "decides" to invade. For that reason, the multicellular organism has a right to defend itself from the virus by expelling or destroying the virus before it can do damage to the organism's internal constitution. But literally every society of actual entities has some value in virtue of the way that it first came into being—that is, through a corporate decision of a set of dynamically interrelated self-constituting subjects of experience—and then continues to exist through repetition or modification of that common element of form in subsequent sets of constituent actual entities. Every society, accordingly, has a determinate character that makes it at least somewhat different from all its contemporaries and thus endows it with objective value in its own right.

At the same time, all this happens only in virtue of divine creativity, which, following Whitehead on this point, is communicated to actual entities through divine "initial aims" with their impact on what Whitehead calls the "subjective aim" or sense of direction for each developing actual entity in its self-constitution.[44] Whitehead himself, to be sure, limited the activity of these divine initial aims to simply providing a directionality to the actual entity in its process of self-constitution and nothing more. But in my view, if Creativity is not a reality different from God but, as noted above, the dynamic nature of God as Creator of heaven and earth, then the divine initial aim not only provides a directionality to the developing, finite actual entity but empowers it to implement its own subjective

---

44. Whitehead, *Process and Reality*, 244.

aim and thus to become what it ultimately decides to be. So Creativity is decisively at work in the self-constitution of each actual entity even when a given actual entity decides not to follow the directionality proposed to it by God on a feeling level. One could compare this understanding of Whiteheadian divine initial aims to the classic Christian understanding of actual grace, that which both empowers and motivates human beings to choose good and avoid evil. The difference, of course, between the two concepts is that the concept of actual grace is limited to creatures that can exercise free will with respect to its acceptance or rejection; thus, actual grace is given to human beings rather than to nonhuman creatures lacking this power of free choice. Following Whitehead on this point, I claim that divine initial aims are given to all actual entities in their moment of self-constitution, even to the actual entities that are constituents of inanimate things. For all actual entities without exception have some limited power of self-constitution and thus in principle can reject or in any case modify the divine initial aim proper to themselves in their moment of self-constitution.

Yet the key point here is that there is still room within this scheme for the classical understanding of God as the transcendent source and inward motivation for the achievement of value on a finite level. Value, in other words, is not totally self-generated by finite actual entities in their moment of self-constitution. Given the influence of God on the developing actual entity through provision of an initial aim for that moment of decision, God still plays an important role in the achievement of value within the cosmic process. Furthermore, the finite value achieved by an actual entity and its successors in a given society can as a result be measured and further evaluated in terms of transcendent values, which are resident in what Whitehead calls the "primordial nature" of God—that is, God's vision of ordered possibilities for both our own and any other finite universe.[45] Yet as Whitehead notes in that same context, God is thereby "not *before* all creation, but *with* all creation."[46] For the de facto achievement of value by finite actual entities at any given moment, God is dependent on the contingent decisions of creatures. Human beings, of course, with their greater power of free self-constitution, are both personally and corporately much more responsible for the achievement of value in this world than nonhuman societies of actual entities. So God especially needs the cooperation of human beings for achievement of divinely intended goals and values for the cosmic process.

45. Ibid., 343.
46. Ibid.

At the same time, in and through what Whitehead calls the "divine consequent nature," all these contingent decisions of finite actual entities are continually ordered and reordered by God into a comprehensive whole, thus preventing total chaos from regularly taking place in the universe and making possible new divine initial aims and new self-constituting decisions of finite actual entities in the future.[47] In my own trinitarian scheme for the God-world relationship, the consequent nature of God basically corresponds to the above-mentioned divine matrix, the all-embracing field of activity proper to the divine persons in their ongoing dynamic interrelation, insofar as it is likewise the field of activity or environment for the world of creation and the evolutionary process. Within this divine matrix as the ultimate source or ontological ground for the power of creativity at work in the world, the evolutionary process produces from moment to moment all the finite actual entities and the societies that they bring into existence through their dynamic interrelation. The divine persons and all their creatures are thus together bringing about what the Christian Bible refers to as the kingdom of God. The kingdom of God is, accordingly, not the work of God alone nor simply the achievement of the finite entities of this world. It is a genuinely cooperative venture between God and finite entities at every moment. Yet contrary to conventional wisdom that suggests that man proposes and God disposes, this is a situation where God proposes and finite actual entities by their self-constitution here and now dispose or decide. The result is a world in which values are often quite ambiguous, a mixture in most cases of both positive and negative components. But the end result, as indicated earlier in this paper, is presumably in the eyes of the triune God an aesthetic, if not always an easily recognized moral, achievement. Something eminently worthwhile is being worked out through the cosmic process even if we human beings, with our strictly limited goals and values, often find it hard to realize and properly appreciate. Our human world, and indeed the entire universe, is still a work in progress.

---

47. Ibid., 350–51.

3

Intelligent Design and the Self-Organization of Nature

Jozef Zycinski, Archbishop of Lublin, Poland, with a double doctorate in theology and the philosophy of nature, is the author of a recent book on religion and science that has been translated into English as *God and Evolution*.[1] In the introduction he makes the following programmatic statement:

> In the still-popular attempts to find a Christian evolutionism, we most often find either appeal to St. Augustine's concept of *rationes seminales* or indications of the moments in which God had to resort to extraordinary interventions in order to introduce new qualities into nature. In the view proposed in this work . . . we will get, not a view of evolution in which the central role will be the classical concept of "plan," "project," and "order," but one in which God, participating in a cosmic *kenosis*, draws to Himself an evolving world, acting as a "Divine Attractor" in situations of chaos, bifurcation, and lack of explicit determination.[2]

While the terminology which Zycinski here employs is not quite the same as that used by Alfred North Whitehead in *Process and Reality*, the vision of God as working in and through the evolutionary processes of nature rather than contravening those processes or temporarily suspending their normal operation is remarkably the same for both men. Likewise, Stuart

---

1. Jozef Zycinski, *God and Evolution: Fundamental Questions of Christian Evolutionism*, trans. Kenneth W. Kemp and Zuzanna Maslanka (Washington, DC: The Catholic University of America Press, 2006).
2. Ibid., 5–6.

Kauffman, whose work *Reinventing the Sacred* we have already reviewed in chapter one, argues on purely naturalistic grounds that classical reductionism or purely "bottom-up" causality within nature must be supplemented by "top-down" causality in the form of an emergent principle of self-organization even at the molecular level of existence and activity. Yet he too stops short of working out a metaphysical cosmology such as Whitehead produced in *Process and Reality* and other works. In this chapter, accordingly, I will first provide a rapid overview of Zycinski's argument for the workings of God in nature as the "Divine Attractor" rather than as an absolute monarch or, in Thomistic terms, the First Efficient Cause. At the same time, I will indicate how Zycinski's cosmological scheme could profit from further specification in terms of Whitehead's metaphysics. Afterward, I will summarize Stuart Kauffman's proposal of an innate principle of self-organization within nature to supplement Darwin's principle of natural selection as the mechanism for the evolutionary process. I will also indicate how my neo-Whiteheadian vision of the God-world relationship both corroborates and to some extent further specifies Kauffman's hypothesis. All this should make clear how God works through natural processes rather than intervenes into those same processes so as to keep the cosmic process moving toward its predetermined goal.

Zycinski on God and Evolution

In part one of his book, Zycinski first takes note of Darwin's own ambivalent feelings about the role of Divine Providence in the process of evolution and then indicates how as a result two highly polarized views about the role of God in evolution came to center stage in the years following Darwin's death: on the one hand, Christian fundamentalism argues that belief in evolution is incompatible with biblical revelation; on the other hand, scientific fundamentalism argues that belief in evolution logically precludes belief in God and the reality of the supernatural and that "nature is all there is." Zycinski ends part one with an appeal to statements of Pope John Paul II and others about the need for "creative dialogue between contemporary biology and a theology free from fundamentalist distortions."[3]

In part two, Zycinski takes on several key issues in the understanding of evolution from both a scientific and a religiously oriented perspective. He distinguishes carefully, for example, between methodological naturalism and ontological naturalism. Methodological naturalism only claims that

3. Ibid., 74.

scientific theories should make no appeal to divine intervention or other "non-physical" factors in the explanation of empirical data.[4] Ontological naturalism flatly denies the existence of God and the reality of the supernatural.[5] Zycinski also distinguishes between teleology and purposiveness within nature. Teleology, as set forth in classical metaphysics, presupposes a conscious agent working toward a predetermined end; it thereby stands in opposition to the laws of "matter-in-motion" formulated by Galileo and Newton at the beginning of the modern era.[6] But teleology is not the same as the purposiveness exhibited by nature on a broader scale (e.g., within systems that exhibit not only moment-by-moment interactions between their parts or members but long-term development toward an end or goal not yet fully specified).[7] What is ultimately at stake here is how one understands the term "laws of nature." Are laws of nature operative the same way in each and every situation, or are they based on statistical laws of probability in which state A, to be sure, imposes with physical necessity state B, but only if and when it excludes at the same time the occurrence of states C, D, and E?[8] Zycinski's conclusion is that in many cases a physical process can be adequately described only in terms of both deterministic and teleological considerations.

Especially within nonlinear thermodynamic systems, the classical distinction between order and chaos becomes more complicated: "States to which a thermodynamic system tends are often called *attractors*, since it is possible to say that, in a certain sense, they draw the evolution of the system to themselves, locally defining its directedness."[9] Zycinski will later describe God in the person of Christ as the Divine Attractor. But here he focuses on purely natural factors within a nonlinear thermodynamic system that move the system as a whole unexpectedly in one direction rather than another. "Part of the energy which was previously used for incoherent thermic motions comes now to be used for the organization of the whole."[10] Order emerges out of apparent chaos in virtue of the inherent dynamism of a nonlinear system far from equilibrium or a so-called "dissipative structure." In principle, then, the movement from nonlife to life within nature can conceivably take place through the operation of

---

4. Ibid., 79–82.

5. Ibid., 81. Zycinski cites here John F. Haught, *God After Darwin: A Theology of Evolution* (Boulder, CO: Westview Press, 2000), 63–64.

6. Zycinski, *God and Evolution*, 98.

7. Ibid., 111.

8. Ibid., 115.

9. Zycinski, *God and Evolution*, 133.

10. Ibid., 135.

purely physical laws, some of which are deterministic in character and others only probabilistic.[11]

In part three of his book, Zycinski sets forth different ways in which God may be conceived in relation to the world of creation. Scripture (the Hebrew and Christian Bible) proclaims that the glory of God is to be found in the works of nature. In both the Middle Ages and the early modern period of Western civilization, God was conceived as the Creator and Sustainer of the laws of nature. But, as Zycinski points out, this presupposes that the laws of nature are deterministic, not probabilistic. What happens by chance is clearly a threat to divine omnipotence unless it can be explained as a miracle or some other form of divine intervention into the workings of nature.[12] Yet given a new scientific understanding of indeterminacy at the quantum level and of stochastic processes bringing order out of apparent chaos at higher levels of organization within nature, a new understanding of God's activity within nature seems to be required: "God is no longer an absolute ruler, forcing a necessary scenario on an evolving nature, but is one of the factors influencing the process of evolutionary transformations."[13] For Zycinski, accordingly, God must be immanent within creation as well as transcendent of it. But how is this to be imagined?

Zycinski indicates that he favors panentheism, the belief that creation exists within God but remains distinct from God in its own mode of existence and activity. Zycinski finds precedent for this view in the writings of the late medieval thinker Nicholas of Cusa. According to Cusa, the created world with its limited set of possibilities represents a "contraction" of the absolute potentiality of God. Thus, God "is the Beginning in which 'everything that in any way either exists or can exist is enfolded. . . . And whatever has been created or will be created is unfolded [emergent] from Him, in whom it is enfolded.'"[14] Zycinski also makes reference to the philosophy of Alfred North Whitehead and other process-oriented contemporary thinkers such as John V. Taylor, Bernard Meland, and Arthur Peacocke in arguing for panentheism as a suitable contemporary model of the God-world relationship.

Yet if God is truly immanent within the evolutionary process, why does that process involve so many dead ends and so much pain and suffering along the way? Zycinski's answer is initially philosophical. Human beings

11. Ibid., 137.
12. Ibid., 152.
13. Ibid., 157.
14. Ibid., 174. Cf. Nicholas of Cusa, *Trialogus de Possest*, 8/19–22.

and other higher-order animal species are sensitive to pain and suffering because they have finely tuned nervous systems capable of great joy as well as great suffering[15] Likewise, in a finite world not all values can be simultaneously or perhaps even successively realized. Such inevitable limitations can also be a source of disappointment and pain to those forced to make hard choices. In the end, however, Zycinski appeals to divine revelation. God's involvement in the cosmic process is marked by kenosis, the self-emptying of God as exhibited by Jesus in his willingness to suffer and die a painful death for the sake of his human brothers and sisters. Likewise, there is the kenosis of the Holy Spirit who keeps alive in human minds and hearts the hope of a brighter tomorrow amid disappointments keenly felt today.[16] Thus, the kenosis of God within the evolutionary process "gives a new dimension to reflection on the cosmic meaning of suffering. The pain does not, thereby, become less, but it receives a radically different meaning."[17]

Finally, in part four of his book, Zycinski reflects on the role of human beings within the cosmic process. The distinctiveness of the human species among the primates is to be found more in cultural differences than in genetics, for example, "the aesthetic sense, abstract thought, creative language, and the moral sense."[18] Zycinski is critical of Christian fundamentalists who fear any linkage of the human species with chimpanzees and other primate species. But he also critiques scientists within the field of sociobiology who interpret evolution so deterministically as to exclude all reference to the spiritual dimension of human life and the activity of God within the evolutionary process.[19] He concludes: "There are no substantive reasons for placing sociobiology and the Christian view of the world in opposition to one another. Both our religious beliefs and our altruism could have genetic foundations without ceasing to express genuine truths and free moral choices."[20]

In a concluding chapter, Zycinski reviews the biblical account of original sin and sees it as descriptive of an important stage in the moral development of the human species. On the one hand, the fall of humankind from its original moral innocence was a step forward in that human beings were forced to take responsibility for their free choices. But on the

15. Ibid., 183.
16. Ibid., 187–88, 191.
17. Ibid., 194.
18. Ibid., 199.
19. Ibid., 211.
20. Ibid., 214.

other hand, the fall closed off an avenue of cultural development in which human beings could have used their freedom more sensibly.[21] The role of Jesus in the world as it presently exists, accordingly, is to be the Divine Attractor, drawing human beings to make free choices more in accord with their deeper instincts and desires, even though these same choices do not always confer immediate benefits in the struggle for existence. Guidelines for such free choices are to be found in the Beatitudes first proclaimed by Jesus in his Sermon on the Mount and then exemplified in his passion and death on the cross.[22] God thus offers human beings not a detailed master plan but a vision for the future, a new directionality for the ongoing evolution of the universe.

Throughout *God and Evolution*, Zycinski makes frequent references to the philosophy of Alfred North Whitehead. He notes with appreciation, for example, Whitehead's resistance to strict determinism in the interpretation of the laws of nature, his description of God as "the subtle Poet of the world, who directs an evolving nature toward His ideals of beauty and goodness,"[23] his prioritizing of future potentiality over current actuality, and finally his conviction that God works in the world through persuasion, not coercion. But as noted above, Zycinski does not systematically adopt for his own use Whitehead's metaphysical scheme nor that of any other contemporary philosopher or theologian. Yet there are problems of consistency with this eclectic approach. For example, if, as Zycinski claims, God works with creatures through a principle of "attraction," not external force or coercion, then logically one should likewise affirm in line with Whitehead's scheme that "the final real things of which the world is made up" are actual entities, momentary self-constituting subjects of experience.[24] For only subjects of experience, not things, can be subject to divine attraction in the form of what Whitehead calls divine "initial aims."[25]

Kauffman on Self-Organization in Nature

In the preface to his book *At Home in the Universe*, Stuart Kauffman challenges the claim that natural selection is the exclusive mechanism

---

21. Ibid., 237.
22. Ibid., 242.
23. Ibid., 163.
24. Alfred North Whitehead, *Process and Reality: An Essay in Cosmology*, corrected edition, ed. David Ray Griffin and Donald W. Sherburne (New York: Free Press, 1978), 18.
25. Ibid., 244.

for biological evolution: "Another source—self-organization—is the root source of order. The order of the biological world, I have come to believe, is not merely tinkered, but arises naturally and spontaneously because of these principles of self-organization—laws of complexity that we are just beginning to uncover and understand."[26] Natural selection, in other words, comes into play only after a certain level of self-organization has been already achieved. At that point natural selection decides which such novel experiments in self-organization will survive and prosper and which, for various reasons (both internal and environmental), will inevitably fail. Thus, only a combination of self-organization and natural selection ultimately explains first the emergence of life from nonlife and then the amazing diversity of biological species that have historically come into existence in the last four billion years on this planet.

Kauffman readily concedes that there is as yet no commonly agreed upon conceptual framework among biologists for conjoining the principle of natural selection with principles of self-organization within nature.[27] But in *At Home in the Universe* and in his later, more technically written book *Investigations*,[28] he sets forth a generalized formula for the way in which self-organization and higher orders of complexity spontaneously appear not only in the life-world but also in economic and political systems. This is a very ambitious project undertaken by Kauffman and his associates at the Santa Fe Institute in New Mexico. In this chapter, I will content myself first with analyzing the general principles for his approach to spontaneous self-organization within nature. Then I will indicate how the new understanding of "societies" within the metaphysics of Alfred North Whitehead that I have sketched in earlier chapters of this book might provide a clue to

---

26. Stuart Kauffman, *At Home in the Universe: The Search for the Laws of Self-Organization and Complexity* (New York: Oxford University Press, 1995), vii. In their comprehensive review of the history of Darwinism, David Depew and Bruce Weber conclude that Kauffman with his theory of self-organizing systems stands midway between two rival research traditions in evolutionary biology, namely, developmentalism and standard Darwinism. Developmentalism emphasizes the inner-driven activity of the organism; standard Darwinism, its relatively passive adaptation to changes in the external environment. See David J. Depew and Bruce H. Weber, *Darwinism Evolving: Systems Dynamics and the Genealogy of Natural Selection* (Cambridge, MA: Massachusetts Institute of Technology Press, 1995), 429–30. They argue that Kauffman's approach might well be the way in which Darwinism itself will evolve to explain more complex patterns of adaptation and change in nature.

27. Kauffman, *At Home in the Universe*, 8.

28. Stuart Kauffman, *Investigations* (New York: Oxford University Press, 2000).

that generalized "conceptual framework" that Kauffman still finds lacking in his own and others' work at present.

In an early chapter of *At Home in the Universe*, for example, Kauffman claims that "life is a natural property of complex chemical systems, that when the number of different kinds of molecules in a chemical soup passes a certain threshold, a self-sustaining network of reactions—an autocatalytic metabolism—will suddenly appear."[29] Let us now compare Kauffman's proposal with Whitehead's definition of a society:

> A nexus enjoys "social order" [is a society] where (i) there is a common element of form illustrated in the definiteness of each of its included actual entities, and (ii) this common element of form arises in each member of the nexus by reason of the conditions imposed upon it by its prehensions of some other members of the nexus, and (iii) these prehensions impose that condition of reproduction by reason of their inclusion of positive feelings of that common form.[30]

For Whitehead, accordingly, each actual entity is a unique subject of experience with its own pattern of existence and activity. Yet all the actual entities within a society still have an analogous self-constitution by reason of their common "prehension" of the pattern proper to the self-constitution of their immediate predecessors in the same society. This "common element of form" carried over from one set of constituent actual entities to another constitutes their group identity as a society.

Structurally, Kauffman's notion of self-organizing chemical systems and Whitehead's notion of a society as a set of actual entities linked by a common element of form seem to be quite similar. Both self-organizing chemical systems for Kauffman and societies for Whitehead are socially organized realities emergent out of the dynamic interplay of their component parts or members. In his book *Investigations*, Kauffman frequently uses the term "autonomous agents" to describe self-organizing systems as entities

---

29. Kauffman, *At Home in the Universe*, 47. As others have noted and as Kauffman himself concedes, he has developed this theory for the emergence of life from the self-organization of molecular components not from observation and experimentation in Nature but from Boolean networks and other mathematical models with computer-generated results. See, for example, Depew and Weber, *Darwinism Evolving*, 431–33; Kauffman, *At Home in the Universe*, 75–86, 99–111. But at this exploratory stage of investigation into the laws of self-organization in Nature, his theories have generated considerable attention and interest among colleagues not only in molecular biology but also in other areas of research, such as economics and politics.

30. Whitehead, *Process and Reality*, 34.

emergent out of the interplay of their component parts or members.[31] But here one must be careful. For when Whitehead discusses "structured societies" or societies composed of subsocieties of actual entities in *Process and Reality*, he does not make clear how such a higher-order structured society came into being in the first place. Whereas Kauffman believes that life spontaneously emerges from the dynamic interaction of nonliving components (chemical systems), Whitehead simply claims that structured societies that are "living" have a "regnant nexus" of entirely living actual entities which is supported by, but still functionally superior to, the other subsocieties of actual entities that are inanimate, nonliving.[32]

What Whitehead thereby leaves unresolved is the question of the origin of this nexus of "entirely living" actual entities. Did it spontaneously emerge out of the dynamic interplay of the various subsocieties of inanimate actual entities? Or was something else required to bring it into existence? Whitehead claims, for example, that "the growth of a complex structured society exemplifies the general purpose pervading nature"[33] and then attributes this growth in complexity or emergence of new order to an increased intensity of experience among the constituent actual entities. But how does this increased intensity of experience arise in the first place? Is it the result of the self-constituting activity of each actual entity taken individually, or is it necessarily an effect of their dynamic interrelation as constitutive members of a society? The problem is that Whitehead attributes agency exclusively to individual actual entities and does not seem to recognize the validity of a higher-order collective agency for the society as a whole so that it can function as a reality in its own right. As a result, Whitehead's system by his own admission is a type of metaphysical atomism: "the ultimate metaphysical truth is atomism."[34] Actual entities "are the final real things of which this world is made up."[35] But if so, it is difficult to see how Whiteheadian societies understood as aggregates of similarly constituted but independently existing inanimate actual entities can bring about a life-form radically different from themselves as individuals.

Thus, if one wishes to claim with Kauffman that life is naturally emergent from the dynamic interplay of nonliving components (chemical systems), then some rethinking of the dynamism at work within a

---

31. Kauffman, *Investigations*, 3–4, 8, 29, 68–73, 105, 120, 128–29, etc.
32. Whitehead, *Process and Reality*, 103.
33. Ibid., 100.
34. Ibid., 35.
35. Ibid., 18.

Whiteheadian structured society is required. I suggest for this purpose some further specifications or modifications of Whitehead's scheme: (a) that there is an agency proper to the various subsocieties within a structured society over and above the agencies of their component actual entities, (b) that this agency functions as the structuring principle or common element of form for the subsociety as such and not simply for its component actual entities in their individual self-constitution, and (c) that the structured society as a whole and all its subsocieties be conceived as a set of interconnected and hierarchically ordered fields of activity.

Given these modifications, a single set of inanimate actual entities within a given subsociety of the overall structured society could by their dynamic interrelation here and now and with gentle prompting from what Whitehead calls a divine initial aim spontaneously generate a dramatically new common element of form for their collective reality as a subsociety within that same larger structured society. Furthermore, if this new common element of form with its potential for higher-order existence and activity is not immediately rejected but rather sustained and supported by the next set of actual entities (and their successors) within that same subsociety, then this subsociety could over time become what Kauffman, as noted earlier in this chapter, called a "self-sustaining network of reactions" or the "autocatalytic metabolism" characteristic of life as opposed to nonlife.

Likewise, this single suitably transformed subsociety could then become the autocatalytic metabolism to enable the structured society as a whole to make the transition from a lower- to a higher-level form of existence and activity and thus by degrees to move from nonlife to life in its overall operation. Key here is the way in which change normally takes place among subsocieties of a larger structured society. For if and when a novel form of existence and activity is introduced within a single subsociety through the collective activity of its constituent actual entities, then the member actual entities of all the other subsocieties have to adjust to what has happened in their midst, either by incorporating the change of pattern into their own individual self-constitution or by rejecting it. If they in some way incorporate this structural novelty into their own pattern of operation, then the structured society as a whole will be transformed and could over time make the transition from nonlife to life. If, however, the constituent actual entities of the other subsocieties do not accept this change of pattern proposed by actual occasions in a coexistent subsociety, then the actual entities of the subsociety in which the change of pattern originated will either regress to their previous pattern of coexistence or break up as a subsociety altogether. The actual entities within the other subsocieties will have equivalently suppressed this novel advance within their midst so as to better preserve the order and directionality received from their predecessors in the structured society as a whole. By way of a concrete example, first think of how human beings respond sometimes positively and sometimes quite negatively to an unexpected proposal for change in their customary mode of operation. Then imagine how subjects of experience at the molecular level of existence and activity might in somewhat similar fashion respond to a new and unexpected pattern of activity in their coexistence as a Whiteheadian society.

In any case, this rethinking of Whitehead's category of society seems to correlate nicely with what Kauffman in *At Home in the Universe* says about the unpredictable way that life emerges from nonlife. He notes, for example, that "life evolves toward a regime that is poised between order and chaos."[36] It is never certain whether life will prevail over nonlife or, if it does prevail, what precise form or structure it will take. "In such a poised world, we must give up the pretense of long-term prediction. . . . Only God can foretell the future."[37] In a similar vein, Whitehead says in *Process and Reality*, "Error is the price which we pay for progress."[38] If the need for order prevails over the desire for novelty within the creative process, then stagnation at a given level of existence and activity ultimately prevails. But the opposite alternative is not simply the case. Unbridled desire for novelty undermines the order requisite for stable existence. In brief, then, a balance between order and novelty within a Whiteheadian structured society is the only way for it to survive and prosper.[39]

Furthermore, given that structured societies are Whitehead's generic term not only for inanimate compounds but also for organisms (plants, animals, human beings), even for supraorganic realities like human communities and physical environments, Whitehead as well as Kauffman seems to be saying that the same basic laws of self-organization are operative everywhere in the cosmic process. Kauffman, for example, compares the explosion of new species at the beginning of the Cambrian era on Earth to the rapid spread of new technologies in modern times and then com-

---

36. Kauffman, *At Home in the Universe*, 26. See also Kauffman, *Investigations*, 22: "Communities of agents will coevolve to an 'edge of chaos' between overrigid and overfluid behavior. . . . Moreover, autonomous agents forever push their way into novelty—molecular, morphological, behavioral, organizational." Some of these experiments in novel self-organization work and others fail. Here is where Darwin's theory of natural selection comes into play in the gradual buildup of complexity within Nature.

37. Kauffman, *At Home in the Universe*, 29.

38. Whitehead, *Process and Reality*, 187.

39. Ibid., 103.

ments that "the parallels are striking, and it seems worthwhile to consider seriously the possibility that the patterns of branching radiation in biological and technological evolution are governed by similar general laws."[40] Given the modifications of Whiteheadian societies that I laid out earlier, this makes perfect sense. The laws governing the aggregation of actual occasions into societies/structured fields of activity are everywhere the same. What happens on the organic and, above all, on the supraorganic or institutional level is only a more complex version of what happens at the inorganic level of atoms and molecules. In every instance, novelty arises within a system/society with the slow growth in complexity of constituent actual entities, provided that the change in the common element of form from a single subsociety within a structured society to the structured society as a whole can be sustained and deepened over time.

One should not, of course, overestimate these and other similarities between Kauffman and Whitehead on the overall pattern of self-organization and the emergence of novelty within nature.[41] Yet if one uses this modified Whiteheadian scheme to support Kauffman's hypothesis of a principle of self-organization within nature, then one can readily explain novelty and development within the evolutionary process without appeal to divine intervention except in the form of divine initial aims, feeling-level "lures" communicated by God to constituent actual entities of a given society for guidance in their self-constitution here and now.[42] According to Whitehead, the constituent actual entities can accept or reject these divine lures, thus allowing for chance and contingency within the evolutionary scheme as stipulated by the principle of natural selection. Yet given the way in which God orders and reorders the "decisions" of actual entities in terms of what Whitehead calls the "divine consequent nature,"[43] divine providence can still be said to be operative within the evolutionary process. There is clear movement toward transcendent goals envisioned by God but in a way that protects the autonomy and integrity of the creature here and now.

---

40. Kauffman, *At Home in the Universe*, 205; *Investigations*, 240–41: "Laws for any biosphere extend, presumably, to laws for any economy. Nor should that be surprising. The economy is based on advantages of trade. But those advantages accrue no more to humans exchanging apples and oranges than to root nodules and fungi exchanging sugar and fixed nitrogen that both make enhanced livings. Thus, economics must partake of the vast creativity of the universe."

41. Cf. Joseph A. Bracken, *Subjectivity, Objectivity, and Intersubjectivity: A New Paradigm for Religion and Science* (West Conshohocken, PA: Templeton Foundation Press, 2009), 138–53.

42. Whitehead, *Process and Reality*, 244.

43. Ibid., 344–51.

In an article for *Theology and Science* some years ago, Howard Van Till made reference to the "formational economy" of the universe, namely, "the set of all of the universe's resources, potentialities and formational capabilities that have ever contributed to the actualization of new physical structures and life forms in the course of its formational history."[44] In his mind, both religiously neutral natural scientists and theistic naturalists like himself should agree on the validity of this presupposition for further dialogue and discussion in the area of religion and science. One does not have to appeal to special divine intervention into the processes of nature to account for the appearance of something genuinely new. Given the resources, potentialities, and capabilities already present within the cosmic process, there is no reason to doubt that there is a natural explanation for everything that happens, even though that natural explanation is here and now unknown. That is, in due time, as a result of careful observation of natural processes and appropriate experimentation with the new empirical data thus gathered, one will come up with at least a plausible natural explanation for the phenomenon in question.

With Van Till's proposal I am in full agreement—with one important proviso. Some of those resources, potentialities, and capabilities of the cosmic process are still not known or at least not fully understood. Natural scientists continue to be surprised by new developments in their chosen areas of research. Hence, one has good reason to believe that what here and now seems utterly miraculous, and thus a clear manifestation of divine intervention into nature, may actually be a prophetic sign of things to come, that is, what will in the providence of God take place quite naturally within the cosmic process. To have predicted, for example, that someday human beings would be able to travel through the air at five hundred miles per hour would have seemed incredible to people living in the Middle Ages, but it is a commonplace experience at the present time. From a theological perspective, one can only conjecture that God is very patient in dealing with creatures, leading them via divine initial aims at every moment to the achievement of goals and values that only God can envision at the present time.

Accordingly, as I shall make clear at length in the next chapter, what classical metaphysics termed divine primary causality over against the secondary causality of creatures can and should be understood as God empowering the creature to make its own self-constituting decision even as God afterward orders it within a broader scheme of things, what Whitehead calls the divine consequent nature and what I prefer to say is the

---

44. Howard Van Till, "Dialogues," *Theology and Science* 2 (2004): 178.

divine field of activity for the three divine persons of the Christian doctrine of the Trinity (in biblical terms, the kingdom of God). Hence, "intelligent design" in the form proposed by Michael Behe, John Dembski, and others to explain alleged "irreducible complexity" within the evolutionary process is not required.[45] This is not, however, to assert that the evolutionary process is purely random, "a directionless process, going nowhere slowly."[46] Intermediate between unchanging divine predetermination and pure randomness is what Pierre Teilhard de Chardin, celebrated author of *The Phenomenon of Man*, termed "directed chance."[47] Given sufficient time and the ongoing persuasive lure of divine initial aims, a set of Whiteheadian actual occasions can naturally evolve to a higher-order level of organization and complexity within nature. As Josef Zycinski makes clear in *God and Evolution*, God acts "as a 'Divine Attractor' in situations of chaos, bifurcation, and lack of explicit determination."[48]

At the same time, one should honestly concede that the movement in the cosmic process toward ever-greater complexity and self-organization is not in itself a proof for the necessary existence of God as transcendent source of the order and presumed directionality of the cosmic process. Quite apart from belief in God, one can from a purely naturalistic perspective claim that the universe is the way it is because of the conjoint operation of bottom-up and top-down causality at all the different levels of existence and activity within nature. My modification of Whitehead's category of society even lends itself to that interpretation. Bottom-up causality is provided by the dynamic interplay of constituent actual entities at every moment of the cosmic process in response to changes in their environment. Top-down causality is provided by the common element of form for the society as a whole thereby generated and afterward preserved in the structure of the society as a relatively stable common field of activity for later generations of constituent actual entities. Everything is perfectly natural; no influence of a transcendent source of meaning and value is strictly required.

One could, of course, counterargue that over the long term, real novelty in the form of new creative possibilities would not likely emerge within a

---

45. Michael J. Behe, *Darwin's Black Box* (New York: Free Press, 1996); William A. Dembski, *The Design Inference: Eliminating Chance through Small Probabilities* (Cambridge: Cambridge University Press, 1998).

46. Michael Ruse, *Can a Darwinian be a Christian? The Relationship between Science and Religion* (Cambridge: Cambridge University Press, 2001), 164.

47. Pierre Teilhard de Chardin, *The Phenomenon of Man*, trans. Bernard Wall (New York: Harper & Row, 1965), 110.

48. Zycinski, *God and Evolution*, 5–6.

conventional bottom-up and top-down interaction between different levels of existence and activity within nature. There is need, accordingly, for a transcendent source of meaning and value to allow something genuinely unexpected to happen. But the naturalistic argument against this proposal would be that the cosmic process itself sooner or later will try out every conceivable possibility. It might take a very long time, but in the end, first the possibility of a rational species like us and then its actual realization would emerge within the cosmic process.

So the ontological naturalists and the theistic naturalists should equally concede that neither of them has a totally compelling proof for either the existence or the nonexistence of God as the Architect of the cosmic process. The universe can be explained either from a purely naturalistic or from a carefully nuanced supernatural perspective, that is, the assumption that the created universe originated and even now continues to exist within the overarching field of activity proper to the three divine persons. Yet even though it exists within the divine field of activity, the universe has its own ontological integrity and follows its own self-generated laws as a very large but still finite network of interrelated fields of activity for constituent actual entities. Hence, neither classical theism (with its sharp distinction between spirit and matter) nor ontological naturalism (with its belief that nature is all there is) but only panentheism (which holds that everything exists in God but is still distinct from God) seems to be a fully satisfactory model for understanding the world around us. In the next chapter I will illustrate this point by discussing both the strengths and the weaknesses of classical theism with its recourse to the distinction between primary and secondary causality for the governance of the cosmic process.

4

Rethinking Primary and Secondary Causality

Without a doubt the notion of evolution has captured the imagination of intelligent people around the world in the years since the publication of *The Origin of* Species by Charles Darwin in 1859. Not only in the field of biology but in all the other natural and social sciences, even in the traditional humanities, the notion of ongoing change and historical development is now more or less taken for granted. There are, to be sure, many contemporary theologians who still strongly affirm the enduring validity of Thomistic metaphysics for contemporary cosmology, especially when it is freshly rethought along the lines of transcendental Thomism, such as that set forth by Karl Rahner in the mid-twentieth century. One of these theologians is Denis Edwards, well known in the area of religion and science, who has set forth in *How God Acts* a persuasive argument for God's working in creation through secondary causes and thus not through direct intervention into the normal operation of the natural order.[1]

In the first part of this chapter, I will offer a summary and analysis of Edwards's argument for a new trinitarian understanding of the God-world relationship that takes into account the contemporary notion of the universe as an evolutionary process. In particular, I will examine Edwards's philosophical presuppositions in making his case and, above all, his reliance on both classical Thomism and the transcendental Thomism of Karl Rahner. Then in the second part of this chapter, I will once again set forth my revision of the process theology of Alfred North Whitehead as another possibility. So in the end, whether one likes Edwards's approach to the God-world relationship or my own revision of Whitehead's, one will have

---

1. Denis Edwards, *How God Acts: Creation, Redemption, and Special Divine Action* (Minneapolis, MN: Fortress, 2010).

picked up an evolutionary understanding of the God-world relationship with a special focus on the doctrine of the Trinity.

Edwards on How God Acts

In an introductory first chapter, Edwards reviews the dominant characteristics of the universe as revealed by natural science. He notes, for example, that the universe appears to have been in ongoing evolution from the big bang onward. Yet it is at the same time a unified totality or ontological whole in virtue of fixed "patterns of relationship" among components at different levels of existence and activity. Each such level has its own integrity even as it serves as a component in a still-higher level of existence and activity. Furthermore, the universe seems to exhibit a surprising movement toward ever-greater organization and complexity through a process of trial and error. What initially appears to be random eventually fits into a definite directionality for the cosmic process as a whole. Yet such a trial-and-error approach to reality has been quite costly in terms of pain and suffering for countless living creatures, with extinction of entire species as the price to be paid for the emergence of higher forms of life, such as our own human species.[2]

Then in chapter two, Edwards reviews what Jesus himself experienced in his relation to God as Father (*Abba*) and conveyed to his contemporaries as God's purpose for them, namely, the establishment of the kingdom of God as a new creation, a new way of living in this world, and the promise of an even better life in the world to come. The various parables of Jesus about the kingdom of God, his healing ministry to those burdened by suffering and pain, his willingness to share food and drink with public sinners and other outcasts of Israelite society, and, finally, his decision to gather a chosen band of disciples and train them in this new way of life—all these features of Jesus' public ministry make clear that Jesus had foremost in his mind the kingdom of God as a reality already present in this world and yet still in the process of development.[3]

Edwards then concludes: "The God who acts in creation, the God who acts in the history of Israel, has now acted in Christ to bring healing and hope to the world in a new creation. What has already begun in Christ will reach its promised fulfillment when all things will be transformed and made new."[4] Thus, God "lovingly waits upon creation" and "suffers with

2. Ibid., 1–14.
3. Ibid., 15–25.
4. Ibid., 25–26.

creation" in its growth through an evolutionary process based on trial and error.[5] But does God's "suffering love" logically imply that human beings can on occasion say no to God's offer of life and love? For that matter, what about all the other creatures on the face of the earth? Do they too have some limited capacity to respond either positively or negatively to God's self-giving love? Edwards is ambivalent on this point. Because he wants to affirm both the unchanging God of classical metaphysics and the suffering God to be found in contemporary process theology, he ends up appealing to divine mystery. God suffers with creation "only in a strictly transcendent and divine way that recognizes the limits of analogical language."[6]

Beginning in chapter three, Edwards sets forth his basic hypothesis: that God is active in the world of creation without violating any known laws of nature. He first says that a theology of creation should begin with the Christ event, namely, Jesus' life, death, and resurrection, followed by the sending of the Spirit to the apostles at Pentecost.[7] In this way, God's self-giving love "can be taken as a proper description of God's action, not only in the incarnation, but also in creation."[8] This is unquestionably an important insight, but here too some ambiguity is present. Edwards seems to move away from the traditional understanding of creation as an expression of divine power, creating something out of nothing by simply willing it to happen. Rather, for him, God's self-giving love in creating the world would seem to involve both gift (divine self-bestowal) and response (acceptance by creatures of God's offer of life and love). This is an important point for Edwards's basic hypothesis. For if the act of creation is based on something like an intersubjective relation between God and creatures rather than on a unilateral exercise of divine power, then there is real contingency as well as necessity within the cosmic process. Intersubjective relations, almost by definition, are grounded in contingent free choices on both sides. But then one can claim that there really is no conflict between a Christian theology of creation and the contemporary scientific hypothesis that cosmic evolution involves both chance events and relatively fixed laws of growth and development. For contingency then plays a key role in both the theological and the scientific understanding of the cosmic process.

---

5. Ibid., 26–33.
6. Ibid., 30. Cf. also on this point Francis J. Caponi, OSA, "Pale Analogies and Dead Metaphors," *Horizons* 37 (2010): 37–42, where he makes clear how Aquinas carefully distinguished metaphor from analogy. For Aquinas, analogy is grounded in metaphysics; metaphor is more the work of a freewheeling imagination.
7. Edwards, *How God Acts*, 35.
8. Ibid., 36.

In chapters four to six, Edwards tackles the key issue of special divine acts: for example, the working of miracles by Jesus during his public life and, in particular, the biggest miracle of all, the resurrection of Jesus on Easter Sunday. How can one claim that God works through secondary causes, the established laws of nature, even in these extraordinary cases? With reference to miracles worked by Jesus during his public ministry, Edwards proposes, quite rightly in my judgment, that miracles should not be understood as suspension of the laws of nature through divine intervention but rather as "wonders of God that take place through natural causes."[9] The challenge, of course, is to explain how that is possible. Edwards's strategy is threefold: first, he insists that the primary causality of God is completely different from secondary causality as found in the natural order; second, he affirms that our present knowledge of the laws of nature is still in the process of evolution and thus radically incomplete; finally, he claims with Karl Rahner that miracles are "signs and manifestations of God's saving action" to people of faith through secondary causes not yet properly understood by modern science.[10] Let us first attend to the distinction between primary causality and secondary causality. Edwards relies here upon Thomas Aquinas, who suggests that the primary causality of God lies in giving existence (*esse*) to creatures: "It is by God's power that every other power acts."[11] But the problem with this line of thought is that God then seems to be responsible for everything that de facto happens in this world, both the bad and the good in equal measure. Edwards, to be sure, claims that God's saving action as revealed in the life, death, and resurrection of Jesus is not meant to eliminate evil but to transform it.[12] This is an excellent answer, but it may need further qualification, as I will indicate later in this chapter.

With respect to Edwards's other two contentions, I have no basic problems. Our human knowledge of the laws of nature is incomplete. Likewise, the miracles of Jesus are, in my judgment, best understood not as expressions of divine power but as eschatological signs of the age to come. Yet I feel uneasy with Edwards's further claim that "the material universe transcends itself in the emergence of life, and life transcends itself in the human."[13] Natural scientists would agree that the cosmic process has in-

---

9. Ibid., 77.
10. Ibid., 77–90.
11. Ibid., 81. Cf. also Thomas Aquinas, *On the Power of God*, trans. English Dominican Fathers (Westminster, MD: Newman, 1952), bk. 3, chap. 7.
12. Edwards, *How God Acts*, 107–27.
13. Ibid.

deed developed from nonlife to life and from life to rational life. But they would also say that this has happened within an evolutionary process based on trial and error. Yet if that is true, then the evolutionary process could have turned out otherwise; quite possibly, there would have been no life and, a fortiori, no rational life, perhaps nothing but subatomic particles in random interaction. So how does the primary causality of God operate to make sure that all of creation is eventually incorporated into the divine life without some form of supernatural intervention into the workings of nature? Edwards is vague on this point.

I conclude this overview of *How God Acts* with a critical evaluation of chapter nine, where Edwards talks about "final fulfillment" or the end of the world. Edwards sets forth his position in terms of two principles: "The first is that the future of our world in God remains radically hidden to us. . . . The second principle is that the future will be the fulfillment of the salvation in Christ that is already given to us."[14] Once again, I readily concur since these principles are foundational to Christian belief. But in line with what Aquinas did with the philosophy of Aristotle in explaining his own Christian beliefs, I also think that one should do more than appeal to the Christ event to justify one's understanding of the goal of the cosmic process. Edwards, however, contents himself with an appeal to the authority of Karl Rahner on this point: "matter will undergo a radical transformation, 'the depths of which we can only sense with fear and trembling in that process which we experience as our death.'"[15] This is, of course, in the end a circular argument. We know nothing about what will happen to us at the moment of death; hence, there is no insight here into the ultimate transformation of matter at the end of the world.

Much more to the point, it seems to me, is his reference to the comments of William Stoeger, a Jesuit priest/scientist, on the relation between matter and spirit:

> We certainly know a great deal about the physical laws that govern the universe. But the more we know about general relativity, particle physics, quantum mechanics, the origins of matter in the early universe, and the nucleosynthesis of elements in stars, the more counterintuitive and mysterious matter becomes. And we are far from understanding the relationship between the ever-changing matter that makes up our bodies, and our personal and interpersonal "I."[16]

---

14. Ibid., 145.
15. Ibid., 155. See also Karl Rahner, "The Festival of the Future of the World," in *Theological Investigations*, vol. 7 (London: Darton, Longman & Todd, 1971), 183.
16. Edwards, *How God Acts*, 154.

Here we seem to have the starting point for a genuinely metaphysical inquiry into the rational grounds for our Christian belief in final fulfillment or the "new creation" (cf. Rev 21:1-4). In what follows, I will offer my own neo-Whiteheadian metaphysical scheme as one way to proceed in this matter.

A Neo-Whiteheadian Version of How God Acts

At the start, however, let me first admit that my metaphysical scheme is only a model or symbolic representation of the God-world relationship. As such, it only explains some features of how God acts in this world, and it neglects others. Accordingly, as Ian Barbour said years ago in talking about the use of models in both theology and the natural sciences,[17] a good model should be taken seriously but not literally. It is not a blueprint, still less a photograph, of the object of study; but it is or at least should be a rationally consistent set of concepts and images to describe a reality that for one reason or another transcends strict empirical verification. There is no way, therefore, to prove that a model is true or false. One can only ask whether it makes better sense than some other model.

Secondly, while I have a high regard for Whitehead's expertise in the areas of both religion and science, I have over the years become quite critical of certain details in his metaphysical system, especially those which seem to run counter to well-established Christian beliefs. In such instances, as I see it, Whitehead's philosophy can and should be modified to accommodate basic Christian doctrines, just as Aquinas modified Aristotle's philosophy to match his religious convictions.

But before engaging in this revision of Whitehead's metaphysics, let us attend to those features of Whitehead's scheme that in my judgment could make a significant difference to Edwards's account of the God-world relationship if properly understood and applied. As already noted, Edwards claims that God's action in creation and redemption is an act of divine self-bestowal but leaves open the question whether some kind of response from creatures is needed in order to be effective. Here is where Whitehead's foundational insight into the nature of physical reality could clear up that ambiguity. That is, whereas contemporary natural science presupposes that the world is ultimately made up of subatomic particles in combination so as to constitute first atoms, then molecules, and in some cases cells or entire organisms, Whitehead claims that each of these subatomic particles

---

17. Ian G. Barbour, *Religion and Science: Historical and Contemporary Issues* (San Francisco: HarperCollins, 1997), 117.

is in fact an aggregate or "society" of momentary subjects of experience linked by a consistent pattern of existence and activity, what he calls "a common element of form."[18] As a result, for Whitehead, everything that exists is either a momentary subject of experience or a society, a rule-governed aggregate of such momentary subjects of experience following one another in rapid succession.

Thus, when Edwards proposes that divine self-bestowal is the deeper reason for the creation and redemption of the world, I would certainly agree. It fits beautifully within a basically Whiteheadian worldview. God is the divine subject of experience, and all God's creatures (even subatomic particles) are in their own way likewise subjects of experience or aggregates of such subjects of experience. Given this basic intersubjective relation between God and creatures, the self-bestowal of God in the act of creation is invariably accompanied by a response, sometimes positive and sometimes negative, to the divine offer. Hence, Edwards's notion of divine self-bestowal is lifted out of the realm of metaphor and anchored within a coherent metaphysical scheme. This is not to claim, of course, that Whitehead's scheme is true, apart from further investigation. But one is dealing with a philosophical argument rather than making a claim simply on the basis of personal conviction. Thus, the claim can be openly debated on purely rational grounds by reflective individuals of different religious backgrounds.

Edwards, of course, can readily counterargue that Whitehead's metaphysical scheme is itself simply a matter of belief. The great bulk of contemporary natural scientists, for example, pay little or no attention to Whitehead's speculative proposals. Since his basic premise that momentary subjects of experience (actual entities) are "the final real things of which the world is made up"[19] cannot be tested and empirically confirmed, they see no usefulness for their purposes in his argument. Their commitment to scientific method with its insistence on mathematical measurement of empirically tested results of experiments makes them very reluctant to engage in such purely philosophical argument. But is measurement of empirical results the only valid criterion for assessing the truth or falsehood of a given proposition? Are there not presuppositions of scientific method, such as the reliability of the principle of cause and effect in predicting the course of future events, which cannot be proven (as David Hume made

---

18. Alfred North Whitehead, *Process and Reality: An Essay in Cosmology*, corrected edition, ed. David Ray Griffin and Donald W. Sherburne (New York: Free Press, 1978), 18, 34–35.

19. Ibid., 18.

clear years ago[20]) but must be taken for granted? For that matter, are there not controversial philosophical issues that remain for natural scientists after determining mathematically how nature de facto works, for example, at the quantum level where both contingency and determinism have to be factored into the formulation of scientific hypotheses? Hence, the dismissal of metaphysical reflection by natural scientists might well be premature, to say the least. As Etienne Gilson commented years ago, metaphysics has a way of burying its undertakers.[21]

In any case, still another feature of Whitehead's metaphysical scheme that might be very useful as solid philosophical backing for Edwards's claims about the God-world relationship is the Whiteheadian notion of creativity, which by definition empowers actual entities as momentary self-constituting subjects of experience to become themselves.[22] Edwards, to be sure, likewise claims that in the act of divine self-bestowal God empowers creatures, above all human beings, to be free to become themselves in relative independence of God.[23] But whereas for Whitehead creativity in its operation is morally neutral, enabling both good and evil to exist in this world, Edwards has to explain how God is responsible for evil as well as good as a result of God's unilateral conferral of existence on creatures. In fairness, Whitehead's own understanding of creativity as a metaphysical principle of empowerment is also somewhat problematic, since in separating creativity from God and treating creativity as a higher-order reality within the cosmic process, God for Whitehead equivalently becomes a "creature" of creativity much like any finite actual entity.[24] But Whitehead still has in his understanding of the relation between God and creativity a provision that could be quite useful for Edwards in buttressing his own argument, previously noted, that God transforms evil rather than totally eliminates it.

Whitehead asserts that God provides "initial aims" to finite actual entities at the beginning of their process of self-constitution, their becoming themselves.[25] So God gives the developing actual entity a directionality or felt sense of purpose in its self-constitution. Hence, while creativity itself is

---

20. Cf. chapter five.
21. Etienne Gilson, *The Unity of Philosophical Experience* (Westminster, MD: Four Courts Press, 1982), 306.
22. Whitehead, *Process and Reality*, 21.
23. Edwards, *How God Acts*, 48. See also Karl Rahner, *Foundations of Christian Faith: An Introduction to the Idea of Christianity*, trans. William V. Dych (New York: Crossroad, 1978), 79.
24. Whitehead, *Process and Reality*, 88.
25. Ibid., 244.

morally neutral in its operation, God is morally committed to the long-term triumph of good over evil within the cosmic process. Yet all this presupposes that God works as subject to subject not only with human beings but with all nonhuman creatures as well. For Whitehead, this is not a problem since he presupposes that the final real things of which this world is made up are actual entities, momentary subjects of experience. For Edwards, on the contrary, it is a problem since within Thomism, even transcendental Thomism, the final real things of which the world is made up seem to be finite entities, objects of the divine act of creation and redemption, rather than subjects of experience in dynamic interrelation with their Creator. For Aquinas, only human beings have such an intersubjective relationship with God. This is, of course, not to discredit Edwards's claim that in virtue of the divine act of self-bestowal inanimate nature transcends itself in life and nonhuman life transcends itself in human life,[26] but only to point out that this claim is based more on subjective religious conviction than on objective philosophical argument. It is not grounded in a coherent metaphysical system in the same way that Aquinas and Whitehead each proceed in their explanation of the God-world relationship.

At one point in his book, Edwards acknowledges that Whiteheadian process theology offers an alternate explanation for the God-world relationship, but he rejects that option "because of my commitment to divine transcendence, among other things."[27] Here I agree with him that Whitehead's understanding of God in *Process and Reality* does not sufficiently guarantee divine transcendence; rather, it seems to make God as dependent on creation as creation is dependent on God.[28] But here is where I feel entitled to rethink and modify some features of Whitehead's philosophy so as to make it more consistent with my own religious beliefs. To be specific, in my view Whitehead did not attend sufficiently to the inherent social dimension of reality with his emphasis on actual entities as the final real things of which the world is made up. For if he had better thought out his category of society so as to distinguish it as a different kind of entity from its components, namely, actual entities as momentary self-constituting subjects of experience, then he would have had at hand the starting point for a new socially oriented metaphysics or worldview that sees reality as basically composed of social systems, that is, communities or environments with individual entities as their necessary parts or members. In this way,

26. Edwards, *How God Acts*, 44.
27. Ibid., 62.
28. Whitehead, *Process and Reality*, 348: "It is as true to say that God creates the World, as that the World creates God."

individual entities, on the one hand, and communities or environments, on the other, intrinsically depend on one another for their very existence and activity and yet maintain their difference from one another. The whole is more than simply the sum of its parts or members.

Whitehead's hypothesis of internal rather than purely external relations between momentary self-constituting subjects of experience lays the groundwork, to be sure, for such a socially oriented worldview. But in stating that the final real things of which the world is made up are individual (actual) entities, he seems to give the communities or environments into which they aggregate a secondary or strictly derivative reality. Yet there are logical problems that arise from that theoretical stance. From a religious perspective, for example, if communities are nothing more than aggregates of individual subjects of experience, then a Christian does not have a rational explanation for belief in God as triune, a community of divine persons who are one God rather than three gods in close association. Likewise, from a philosophical perspective, one cannot make rationally plausible Edwards's faith claim in *How God Acts* that the triune God offers human beings, and indeed all other creatures of this world (to some extent), salvation and redemption in terms of incorporation into the divine communitarian life. One finds oneself affirming either that the world is absorbed into the reality of God (pantheism) or that "God" is just another name for the overall reality of the world (materialism). So establishing that communities and natural environments are more than simply aggregates of their component parts or members but are instead a different kind of objective reality is crucial for rational acceptance of the doctrine of the Trinity and for Christian belief that we do indeed participate in the divine communitarian life after the death of the body and the end of life in this world.

In a subsequent chapter, I will deal more in detail with Christian belief in life after death and the resurrection of the body from the perspective of this neo-Whiteheadian God-world relationship. For the moment, I simply recapitulate what I have already set forth in this book. I propose that Whiteheadian societies, wherever they are found, are not aggregates of similarly constituted actual entities but rather systems, enduring structured fields of activity for those same actual entities in their dynamic interrelation at any given moment. A field, in other words, is by definition a semipermanent reality that endures the passage of time with a pregiven structure much like a thing or Aristotelian substance; but unlike an Aristotelian substance, the structure of that field is still subject to gradual modification over time. The ever-changing pattern of events taking place within the field will in due time modify the structure of the field. Furthermore, if we apply this notion of a Whiteheadian society as a structured field of activity for the events taking place within it to the analysis of human communities and natural environments, we can argue, first, that human communities like the family, the local community, the nation state, and even the international community as a whole are enduring structured fields of activity for the activities of all the people who live, die, work, and play within them. Secondly, a natural environment can be seen as an enduring structured field of activity for all the living and nonliving things within it in their dynamic interrelation. Finally, if those of us who are Christian believe that God is a community of divine persons, we can logically conclude that the unity of God precisely as a community of divine persons is also a shared field of activity that the divine persons coconstitute in and through their ongoing relationship with one another and with all their creatures.

Given this notion of a field as a different kind of objective reality than a traditional Aristotelian substance, we can then propose that fields, wherever they are found, horizontally overlap so as to have a joint impact on all the people and things contained in each of the fields. Virtually all human beings, for example, belong to more than one community (e.g., one's family, one's fellow workers, those with whom one recreates or worships, etc.) and are strongly influenced by those other human beings, as they together coconstitute structured fields of activity for their common interaction. Likewise, we can picture fields as vertically layered vis-à-vis one another. In this way, smaller fields (such as the field of activity proper to an atom) provide the infrastructure for larger and more complex fields of activity (like that of a molecule or living cell), and the larger fields of activity in turn constitute the law-like context or superstructure for the proper functioning of the lower-level fields of activity. Finally, given the plausibility of this overall field-oriented approach to reality, we can say that the enormous but still finite field of activity proper to creation as a whole is actually contained within the infinite field of activity proper to the divine persons in their ongoing relationship to one another and to us, their creatures.

If all this is accepted as logically plausible, we have at our disposal a philosophical scheme that nicely undergirds Edwards's faith-based proposal that the ultimate purpose and goal of creation is incorporation into the riches of the divine life. That is, if the field of activity proper to creation as a whole is contained within the all-encompassing field of activity proper to the divine persons, then even now we as God's creatures "live and move and have our being" in God (Acts 17:28), albeit with the expectation that at the moment of death we will be more fully incorporated into the divine communitarian life. Moreover, this model or symbolic representation of the God-world relationship seems to protect both God's transcendence of the world and the world's partial independence of God. The divine persons equivalently make "space" within their own divine field of activity for the world of creation together with all its subfields of activity corresponding to particular communities and natural environments. Yet creation as a whole and all the communities and environments, persons and things within creation still possess some measure of autonomy before God, since everything and everyone exists within their own limited field of activity as well as within the all-encompassing divine field of activity.

To conclude, let me repeat that this is only an imperfect model of the God-world relationship. Yet it is basically grounded in the process-oriented philosophy of Alfred North Whitehead and thus provides a measure of rational plausibility for long-standing Christian beliefs about our human relationship to God, one another, and the world of nature. That is, it sets forth a metaphysical scheme that at least in principle is applicable to the natural world as well as to the supernatural order. At the same time, it does not call into question Denis Edwards's conclusions about how God acts in this world but only his argument for arriving at those same conclusions. In the end, of course, rational argument is necessarily subordinate to divine revelation. But as noted earlier in this chapter, while there are inevitably many models, multiple philosophical explanations, for Christian belief in the God-world relationship, some are more rationally convincing than others, if only because they make less appeal to divine mystery, the incomprehensibility of God, as part of the argument.

Part Two

Systems Thinking in the Social Sciences

5

From Platonic Forms to Open-Ended Systems: The Search for Truth and Objectivity

Philosophers and theologians in Western civilization have always sought to determine what is objective and true in the world around them. One sees this passion for objective certitude, for example, in Plato's celebrated analogy of the cave in the *Republic* in which he asserts the priority of the Forms or transcendental Ideas over the confusing data of common sense experience.[1] Truth and objectivity are to be found in the unchanging world of ideas, not in the ever-changing phenomena of sense experience. Aristotle in his *Metaphysics* continued this line of thought with his claim that physical entities are composed of form and matter, with the unchanging substantial form serving as the principle of order and intelligibility for all the contingent material attributes of the entity in question.[2] Medieval theologians, notably Thomas Aquinas in his multivolume *Summa Theologiae*, likewise began their systematic reflections on life in this world with a set of a priori rational arguments for the existence of God the Creator who from all eternity orders everything in heaven and earth to its predetermined end.

---

1. Plato, *The Republic*, trans. Francis MacDonald Cornford (New York: Oxford University Press, 1962), bk. VI, 509D–511B; bk. VII, 514A–521B. Further discussion of Plato's philosophy and of all the other philosophers mentioned in the first half of this chapter can be found in my recently published book *Subjectivity, Objectivity, and Intersubjectivity: A New Paradigm for Religion and Science* (West Conshohocken, PA: Templeton Foundation Press, 2009), chapters 1–7.

2. Aristotle, *Metaphysics*, in *The Works of Aristotle*, ed. W. D. Ross, vol. 8 (Oxford: Clarendon Press, 1928), 1031a–1032a.

Aquinas, for example, at the very beginning of the *Summa* claimed that God's essence or nature is identical with his act of existence.[3] But this is to confuse existence in the conceptual order (the highest and most comprehensive perfection) with existence in the real order of things (that which happens to be the case here and now). That is, by not distinguishing between God as the Supreme Being and Being as what is in fact the case here and now, one risks the charge of pantheism; God is everything and everything is God. Aquinas solved this problem, to be sure, by claiming that creatures participate in God's unlimited or infinite act of existence by exercising their own more limited or finite act of existence.[4] But the logical consequence of this line of thought is that Aquinas's thought system is grounded in religiously based faith claims, which are not empirically verifiable. How, for example, can one prove or disprove on empirical grounds, first, that every material being is a combination of matter and form and, second, that God and angels are immaterial realities with God alone fully in act by nature? This is unquestionably an interesting philosophical argument, but there are still other conceptual alternatives: for example, that matter is an illusion and only spirit exists (as in certain forms of Hindu thought) or that only matter exists with its own intrinsic principle of self-organization (as in the Buddhist notion of dependent co-origination). Moreover, if one objects that at the very beginning of the *Summa Theologiae* Aquinas sets forth five empirically grounded proofs for the existence of God as in different ways the Uncaused Cause of everything else that exists,[5] one should remember that Aquinas simply assumes that this Uncaused Cause is the God of biblical revelation. Likewise, he presupposes the Aristotelian understanding of cause and effect relations whereby the cause necessarily exists prior to bringing into existence the effect. But this philosophical claim can be challenged both by the Buddhist understanding of dependent co-origination and, as we shall see in the next few chapters, by the assumptions of contemporary systems theory. That is, within complex systems it is often very difficult to isolate one-on-one, cause-and-effect relations in the functioning of the system; much as with Buddhist dependent co-origination, everything seems to condition the existence and activity of everything else.

Aquinas, of course, was not the only major Western thinker to engage in this type of implicitly a priori philosophical reflection, nor was he the

---

3. Thomas Aquinas, *Summa Theologica*, trans. Fathers of the English Dominican Province (New York: Benzinger, 1948), I, q. 3, a. 4.

4. Ibid., q. 4, a. 2, resp.

5. Ibid., q. 2, a. 3, resp.

last. Rene Descartes, for example, recognized the overly theoretical character of late medieval scholasticism and tried to remedy this situation by providing his philosophy with an undeniable empirical starting point: "I think, therefore I am" (*Cogito, ergo sum*).[6] But then he argued on purely conceptual grounds for the existence of God as the necessary source of the notion of infinity in his own finite mind and for the reality of the world around him as guaranteed by the veracity of God as the Creator of heaven and earth. His philosophical system is, accordingly, heavily a priori (apart from its empirical starting point).[7] In even more rigorously logico-deductive fashion, Benedict Spinoza argued in his *Ethics* that "the order and connection of ideas is the same as the order and connection of things."[8] But this is to equate empirical cause-and-effect relations with logical ground-consequent relations within his thought system. Thus, the ultimate physical cause of all that exists in nature is at the same time the theoretical first principle of his system, namely, God understood as Substance, "that which is in itself and conceived through itself."[9] The fact that this line of thought reduces all the persons and things of this world to contingent modifications of God as the sole subsistent reality did not apparently weigh heavily with Spinoza since he was intent on working out the logical consequences of his thought system in virtual independence of empirical reality all around him.

Across the English Channel, the British empiricists John Locke and David Hume adopted another philosophical methodology, namely, strict conformity to the data of sense experience. But this strategy unhappily led to the virtual disappearance of logical claims to truth and objectivity in the world around oneself. Locke, for example, used what he called the "historical, plain method" of introspection into the workings of his own mind on the data of sense experience.[10] But insofar as he limited himself to reflection simply on the data of sense experience, he was drawn to adopt uncritically Descartes's presupposition that all we know is our own ideas

---

6. Rene Descartes, *Meditations on First Philosophy*, bk. II, in *The Philosophical Works of Descartes*, 2 vols., trans. Elizabeth S. Haldane and G. R. T. Ross (Cambridge, UK: Cambridge University Press, 1978), I, 150.

7. See Joseph A. Bracken, *Subjectivity, Objectivity, and Intersubjectivity: A New Paradigm for Religion and Science* (West Conshohocken, PA: Templeton Foundation Press, 2009), 28–31.

8. *Spinoza's Ethics and De Intellectus Emendatione*, trans. Andrew Boyle (London: J. M. Dent, 1959), 41 (II, prop. 7).

9. Ibid., 1 (I, prop. 3).

10. John Locke, *An Essay Concerning Human Understanding*, ed. Peter Nidditch (Oxford: Clarendon Press, 1975), 44.

understood as mental representations of extramental things.[11] This led Locke to distinguish between primary and secondary qualities, that is, those which are genuinely representative of innate qualities in extramental things (solidity, extension, figure, motion, or rest) and those which simply reflect the way in which the mind receives and interprets those same sense impressions (colors, sounds, tastes, etc.).[12] From there it was an easy inference to distinguish between nominal and real essences, those which are products of the human mind and those which de facto exist in extramental things. Roughly one hundred years later, David Hume completed the job of undermining claims to truth and objectivity in the natural order when in his *Treatise of Human Nature* he described sense data as "a heap or collection of different perceptions united together by certain relations" but without any necessary reference to the self as a reality distinct from its individual perceptions.[13] Locke had maintained the ontological reality of the Self as the organizing principle for its own sense impressions and ideas. But with Hume's denial of the enduring reality of the Self in sense experience, claims to objective truth lack any empirical grounding. One is lost amid the rapid succession of sense experiences without any way of being able to reduce them to order.

On the European continent, Immanuel Kant had already been suspicious about the validity of traditional metaphysics when he read Hume's *Treatise of Human Nature*. He fully recognized the abstract and largely a priori character of the philosophy of Christian Wolff, which in his earlier years he had so much admired, and was at the same time alarmed by Hume's claim that the law of cause and effect in nature was nothing more than a subjective habit or mental disposition arising from the observation of a regular succession of the same sense impressions in one's consciousness.[14] If Hume were correct, then the pursuit of universally binding laws in the workings of nature was foolhardy. So with his "Copernican Revolution," Kant put forth the claim that philosophical presuppositions that are key to the pursuit of natural science, namely, the principle of causality, the notion of permanence in time (substance), and the experience of reciprocity or community between perceived entities, did not derive from prior sense experience of the external world but were present in human

---

11. Descartes, *Meditations on First Philosophy*, bk. III, in *The Philosophical Works of Descartes*, I, 160–61; Locke, *An Essay Concerning Human Understanding*, 538.

12. Locke, *An Essay Concerning Human Understanding*, 134–43.

13. David Hume, *A Treatise of Human Nature*, ed. I. A. Selby-Biggs (Oxford: Clarendon Press, 1967), 207–8.

14. See Bracken, *Subjectivity, Objectivity, and Intersubjectivity*, 54–55.

consciousness in virtue of the antecedent organizing activity of the human mind upon the data of sense experience.

The human mind, in other words, does not conform to already existing structures in the external world, but the world is perceived and known by the mind in virtue of its own inbuilt structures.[15] In this way Kant restored the possibility of objective truth by stipulating the reality of what he called the transcendental Self, namely, that which manifests itself indirectly to human consciousness in self-awareness or the implicit "I think" accompanying knowledge of everything else.[16] As such, it makes possible valid scientific knowledge of nature based on a combination of empirical data and the organizing activity of the human mind. To the present day, moreover, this is generally the way that scientific method is understood and practiced by many, if not most, natural scientists and, with some modifications, also by most social scientists.

Yet one should remember that Kant paid a heavy price for this restoration of the claim to truth and objectivity in the natural order. For he restricted the scope of truth and objectivity to the contents of his own sense experience and had to deny the possibility of valid knowledge of the natural world apart from the organizing activity of the Self. This implicit dualism between the human spirit and the external world around it represented a great challenge to Fichte, Schelling, and Hegel, the so-called German Idealists. The gap between spirit and matter had to be bridged by making Absolute Spirit the transcendent source of the material world and the guiding force in nature's ongoing mode of operation. Hegel, in particular, achieved this speculative tour de force with his *Phenomenology of Mind*, published in 1807, where he began with reflection on common sense experience, in which the objects of knowledge were thought to be really distinct from oneself, and ended with absolute knowledge, a systematic philosophical understanding of reality from the perspective of Absolute Spirit. Everything in the world is thus only in external appearance a reality unto itself; in fact it is a partial manifestation of Absolute Spirit in progressively revealing itself in nature and in human consciousness.[17] Schelling, who outlived Hegel by more than twenty years, eventually saw

---

15. "Preface to Second Edition," *Immanuel Kant's Critique of Pure Reason*, trans. Norman Kemp Smith (New York: St. Martin's Press, 1964), B xvi. NB: The letters A and B correspond to the text in the first and second editions, respectively, of the *Critique of Pure Reason*.

16. Ibid., B 132.

17. Bracken, *Subjectivity, Objectivity, and Intersubjectivity*, 82. See also G. W. F. Hegel, *Phänomenologie des Geistes*, ed. Johannes Hoffmeister (Hamburg: Felix Meiner, 1952).

through the heavily a priori character of Hegel's philosophy and, for that matter, his own earlier philosophy. So in his lectures on the philosophy of revelation in the 1850s, Schelling insisted that speculative philosophy must have a starting point in an empirical reality outside the thought system itself. With respect to the philosophy of revelation, this extramental reality must be God as a personal Other who freely chooses to create the world and reveal himself through the cosmic process.[18]

Søren Kierkegaard attended Schelling's lectures on the philosophy of revelation at the University of Berlin in the 1850s but was still uneasy with the elaborate thought systems of the German Idealists in which a succession of abstract concepts illustrating the dialectical movement from thesis to antithesis to synthesis played such a key role. For these speculative schemes offered nothing of any value to concrete individuals like him in dealing with the ups and downs of life. So in a series of thinly disguised autobiographical accounts, he sketched his own itinerary of the human mind and heart toward God.[19] He described in graphic terms the move from a self-centered, purely aesthetic approach to life, to a more idealistic but still impersonal ethical stage of intellectual development in which one is motivated by a sense of duty, and from there to the intensely personal religious stage of development in which one acts in response to a special call from God and can only hope that the call is genuine and not illusory.[20] Likewise, in the *Concluding Unscientific Postscript to Philosophical Fragments*, Kierkegaard asked himself whether objectivity or subjectivity is ultimately more important for assessing the truth of Christianity. He concluded that subjectivity is more important since the issue is not about the truth of Christianity but about the individual's relation to Christianity.[21] Kierkegaard, accordingly, initiated a move from truth and objectivity in the form of an abstract and impersonal philosophical system to a new priority of the truth to be found in human subjectivity and ordinary experience. Furthermore, what began somewhat tentatively with the largely autobiographical writings of Kierkegaard in the mid-nineteenth century has now become a recurrent

---

18. Bracken, *Subjectivity, Objectivity, and Intersubjectivity*, 86–88. See also *Schellings Werke*, ed. Manfred Schröter, vol. 11 (Munich: C. H. Beck, 1968), 553–72, esp. 566.

19. Bracken, *Subjectivity, Objectivity, and Intersubjectivity*, 90–95.

20. Søren Kierkegaard, *Fear and Trembling*, trans. Howard V. Hong and Edna H. Hong (Princeton, NJ: Princeton University Press, 1983); likewise *Either-Or*, 2 vols., trans. Howard V. Hong and Edna H. Hong (Princeton, NJ: Princeton University Press, 1987).

21. Søren Kierkegard, *Concluding Unscientific Postscript to Philosophical Fragments*, 2 vols., trans. Howard V. Hong and Edna H. Hong (Princeton, NJ: Princeton University Press, 1992).

theme in the work of many scholars in philosophy, theology, and the other humanities in the twentieth and twenty-first centuries.

To name just a few of those who are profoundly suspicious of a priori metaphysical schemes and other nonempirical attempts at understanding the self, the world, and God, Martin Heidegger in his celebrated book *Being and Time* claimed that in classical Western philosophy, beginning with Plato and Aristotle, ontology, the study of Being, was confused with God as the Supreme Being. Hence, ontology is really onto-theology.[22] About the same time in France, Emmanuel Levinas in his book *Totality and Infinity* critiqued the notion of totality in a self-enclosed philosophical system such as Hegel's since it subtly or even overtly subsumes the individual person or thing into the logic of an overarching conceptual system that alone gives meaning and value to the entities contained within it.[23] Instead, genuine self-transcendence is to be found in the notion of infinity as revealed in the "face" of the Other, the subjectivity or interiority of the Other that cannot be fully comprehended in rational reflection.[24] One of the key French postmodernists, Jean-Francois Lyotard, coined the term "meta-narrative" or grand narrative in order to claim that efforts to provide a comprehensive overview of human history are ultimately illusory.[25] Such "meta-narratives," as Levinas proposed earlier, are in reality thinly disguised attempts to control the interpretation of historical events in one's own favor. Finally, the celebrated French deconstructionist, Jacques Derrida, in his widely read essay "Différance," claimed that the meaning of words in speaking and writing is never settled or complete but always "deferred" through reference to still other words by way of explanation.[26]

---

22. Martin Heidegger, *Being and Time*, trans. John Macquarrie and Edward Robinson (New York: Harper & Row, 1962), 19–35.

23. Emmanuel Levinas, *Totality and Infinity: An Essay in Exteriority*, trans. Alphonso Lingis (Pittsburgh, PA: Duquesne University Press, 1969), 21–30.

24. Ibid., 51.

25. Jean-Francois Lyotard, *The Postmodern Condition: A Report on Knowledge*, trans. Geoff Bennington and Brian Massumi (Minneapolis, MN: University of Minnesota Press, 1984), xxiii–xxiv.

26. Jacques Derrida, "Différance," in *Margins of Philosophy*, trans. Alan Bass (Chicago: University of Chicago Press, 1982), 3–27. See also my book *The One in the Many: A Contemporary Reconstruction of the God-World Relationship* (Grand Rapids, MI: Eerdmans, 2001), 81–94, where I propose that "différence" is actually a metaphysical concept much akin to Whitehead's notion of creativity, that is, a principle of spontaneity whereby something new continually happens but always in the context of an ordered series of events. Hence, like Whiteheadian creativity, "différence" for Derrida implies both constant change and structural continuity at the same time.

As a result, language is always fluid so that the sense of objective truth in human communication is illusory.

In the humanities, all this criticism of the classical understanding of truth and objectivity has been generally accepted as legitimate, albeit with some key reservations. For example, one should be aware of one's "social location," that is, one's family and societal background, in coming to conclusions about what is in fact the case or what ought to be done to change the present situation. To remedy this potential blindness to one's purely subjective inclinations and desires, dialogue with others about issues of common concern is highly recommended. Through the give-and-take of dialogue with other people, one gradually comes to see the inevitable limits of one's customary perspective on life. In listening carefully to the views of other people, especially those who come from a different cultural background, one comes to recognize one's unconscious biases and prejudices in a way that would be virtually impossible simply through extended self-reflection.[27] Yet the practice of dialogue would be fruitless unless the parties involved did find areas of mutual agreement and a consensus about appropriate common action in the light of what appears to be the case. Hence, truth and objectivity still have a place within a context of ongoing dialogue. But what is meant by truth and objectivity in this new situation undergoes a subtle change.

Whereas in the past truth was basically seen as unchanging, even eternal, if somehow identified with God as its ultimate source, truth is now viewed as necessarily processive, subject to gradual change in view of ongoing dialogue with others and altered environmental conditions. Truth, after all, should reflect the way things are, not simply the way that one personally thinks they should be. Truth, accordingly, is more an ideal to be striven for than an accomplished fact. What Mahatma Gandhi called satyagraha, the pursuit of truth, should motivate all conscientious human beings in their quest for meaning and value in life.[28] At the same time, this is not to concede that truth is simply relative to the subjective perception of the individual observer or to the concrete details of the situation at hand. Rather, one should say that truth is necessarily relational, dependent for its objectivity upon a wide variety of factors that need to be taken into account when attempting an objective judgment about a given state of affairs. Such factors include, of course, the views of others involved in the situation at hand and the possibility that what used to be the case is no longer true. With the passage of time, the situation may have changed in

---

27. Joseph A. Bracken, *Christianity and Process Thought: Spirituality for a Changing World* (Philadelphia, PA: Templeton Foundation Press, 2006), 67.
28. Ibid., 65.

ways that no one could have anticipated. In a world still in the process of growth and development, human beings must adjust their thinking and behavior to meet the challenges of a new situation or face a future fraught with danger to personal and group survival. Some truths and values, to be sure, are virtually timeless, valid in almost every conceivable situation. But others are just as clearly time bound, limited by constraints that cannot be set aside without significant negative consequences. It is better to conclude, then, that truth is both processive and communitarian, the result of sustained dialogue between those involved in a given situation with an eye to an effective communal response to whatever needs to be changed. The day that truth was seen as timeless and unchanging, simply a matter of momentary reflection on the part of any reasonably intelligent human being, has long ago ceased to exist, if indeed it ever existed except in the minds of those already convinced of the rectitude of their own opinions.

In the natural and social sciences, truth is likewise considered to be processive and communitarian, but in ways that are different from the processive and communal character of truth and objectivity in the humanities. For in the humanities the governing methodology is generally hermeneutics; the discovery of truth and objectivity is to be found in the ongoing comparison and contrast of varying interpretations of a received text. Where there is a convergence of opinion on the meaning of a text, its truth appears self-evident. Where there is a clash of views on the meaning of a text, a judgment about the truth value and objective meaning of a text has to be deferred until a stronger consensus of opinion emerges as to what the text actually says and what it does not say. Within the natural and social sciences, however, the processive and communitarian character of the search for truth and objectivity takes the form of formulating a hypothesis so as to solve a hitherto controversial issue, testing the theory for its empirical verifiability, and then writing up one's hypothesis and empirical results for publication in a scientific journal. If other scientists then perform the same experiment and achieve basically the same results, one can lay claim to having achieved at least a provisional truth and objectivity in the matter at hand. Yet one fully anticipates that at a later date another simpler theory with even more accurate empirical results will be conceived and empirically tested so as to win the approval of one's peers in the scientific community.

Still another test of truth and objectivity in the natural and social sciences is, of course, predictability and repeatability. One-time events are not normally the subject of scientific investigation. Something has to happen in basically the same way several times in a row to warrant close scrutiny so as first to find out how it happened and then to establish a mathematical formula for determining when it will happen next. So there is a degree of precision in the sciences that cannot be achieved in the humanities, where one is more preoccupied with one-time events, the historical reasons for their occurrence, and the actual or projected future consequences of those same events. There is, accordingly, greater room for speculation in the humanities, but even here one's interpretation of a text or a historical event demands some form of empirical verification/confirmation for others to take one's theory seriously. Finally, whether in the humanities or in the sciences, internal consistency and logical coherence are prerequisite for any satisfactory theory. The details of the system cannot be in contradiction with one another but must reinforce one another so as in the end to constitute a systematically organized whole. As Alfred North Whitehead commented about philosophical schemes (including his own) in the introduction to *Process and Reality*, "Speculative Philosophy is the endeavor to frame a coherent, logical, necessary system of general ideas in terms of which every element of our experience can be interpreted. By this notion of 'interpretation' I mean that everything of which we are conscious as enjoyed, perceived, willed or thought, shall have the character of a particular instance of the general scheme."[29]

So contrary to the protestations of postmodern thinkers about the evils of "totalizing systems," some form of systematically organized thought is indispensable for serious research and publication in both the humanities and the sciences. But granted the need for logically coherent thought systems, what kind of systems should be ambitioned? Here the reservations of the postmodernists should be borne in mind. Totalizing systems or closed systems, such as that of Hegel, that do not allow for renovation or restructuring in the light of unexpected developments in the empirical reality under investigation should be avoided as logical dead ends. As Schelling noted with respect to his own and Hegel's philosophical schemes, totalizing or closed thought systems exist simply in the conceptual order as masterpieces of logical thinking, but they have little or no value to interpret life as it is actually lived. An open-ended system, on the contrary, by definition demands a sense of contingency both in the mind of the system builder (e.g., willingness to revise one's basic presuppositions in the light of new empirical data) and in the object(s) under investigation (e.g., in the life sciences, complex organisms with an inbuilt principle of spontaneity or self-organization). As we shall see in the next three chapters of this book,

---

29. Alfred North Whitehead, *Process and Reality: An Essay in Cosmology*, corrected edition, ed. David Ray Griffin and Donald W. Sherburne (New York: Free Press, 1978), 3.

Whitehead's cosmology or systematic understanding of the God-world relationship does have an inbuilt principle of contingency or spontaneity, given his presupposition that "the final real things of which the world is made up" are actual entities / actual occasions, momentary self-constituting subjects of experience.[30] Unlike material atoms, which are simply tiny bits of lifeless matter, these "spiritual atoms," even at the quantum level of existence and activity within nature, are able to respond in some minimal fashion to the world around them through some unconscious self-constituting "decision." This power of spontaneous self-generation, although heavily conditioned by the environment or structured field of activity in which they arise, allows groups or "societies" of such spiritual atoms to achieve over time progressively higher and more complex patterns of interaction. In this way subatomic particles aggregate into atoms, atoms into molecules, more complex molecules into primitive cells, cells into organisms, and organisms into highly sophisticated environments or communities.

Whitehead terms these progressively more complex fields of activity for constituent actual entities "societies," but one can also refer to these complex groupings of component parts as systems. Ervin Laszlo, for example, in his book *Introduction to Systems Philosophy*, maintains that the basic units of reality are systems, whether natural (the results of cosmic evolution) or artificial (man-made things such as machines or other tools for human use).[31] Artificial systems or things upon closer examination, however, are made up of natural systems (e.g., wood, stone, metal) redesigned by human beings for their own purposes. Natural systems are thus the basic units of physical reality, and Laszlo defines them as follows: "a nonrandom accumulation of matter-energy in a region of physical spacetime, which is nonrandomly organized into coacting interrelated subsystems or components."[32] "Nonrandom" is the key word here, indicating that a system for Laszlo is roughly equivalent to a Whiteheadian society or grouping of self-constituting subjects of experience with a "common element of form."[33] The only real difference between Laszlo's notion of system and a Whiteheadian society is that for Laszlo systems are all that exist; in the end there are only systems and subsystems ad infinitum. For Whitehead, however, while there are indeed "structured societies," societies composed

---

30. Ibid., 18.
31. Ervin Laszlo, *Introduction to Systems Philosophy: Toward a New Paradigm of Contemporary Thought* (London: Gordon and Breach, 1972), 30.
32. Ibid.
33. Whitehead, *Process and Reality*, 34.

of subsocieties,[34] the ultimate components of any society, whether structured or not, are not still other societies (as with Laszlo's notion of systems and subsystems) but actual entities, momentary self-constituting subjects of experience. So within Whiteheadian cosmology, societies are the objective result or necessary byproduct of the interaction of subjective components, namely, actual entities.

This is not a minor point. Laszlo, in saying that systems, whether large or small, are the ultimate units of reality, clearly implies that systems exercise agency in their own right; they are able to combine into higher-order or more complex systems as circumstances demand. Whitehead, however, insists that societies have no agency in their own right. All the agency to be found within a Whiteheadian society is the agency of its subjective component parts, actual entities as momentary subjects of experience.[35] Who is right on this point? In my judgment Whitehead has thought more deeply about the workings of bottom-up and top-down causality within nature. Bottom-up causality is an instance of what Aristotle calls efficient causality, that which makes something happen. Top-down causality, on the contrary, is an instance of final causality for Aristotle, that which objectively guides or orients the collective action of the interrelated agents of change. Bottom-up causality, on the one hand, demands subjectivity since only subjects of existence and activity can be the agents of change. Top-down causality, on the other hand, presumes objectivity, that which structures subjective agency in one direction rather than in another. Top-down causality is thus equivalent to what natural scientists working in biochemistry term "information" rather than "energy." The purpose of information in the DNA of a cell, for example, is not to add new energy to the combination of molecules making up the cell but to organize the molecules into an organized whole, a cell as a unitary, objective reality.

Common sense, to be sure, would seem to say that Laszlo is right and Whitehead is mistaken. After all, if a human being is a complex system made up of subsystems (e.g., the heart, the lungs, the limbs, and, above all, the brain) and if a human being clearly exercises agency in the direction of his or her life, then systems possess agency in their own right. Yet Whitehead would argue that a human being is a structured society made up of multiple subsocieties, each with its own set of constituent actual entities or momentary subjects of experience. So all the agency in a human being from moment to moment is derivative from the collective agency of all the actual entities at work within the mind and body. From moment

---

34. Ibid., 99.
35. Ibid., 34, 89.

to moment I am the byproduct of everything going on inside of me (heart pumping, lungs breathing, limbs moving, mind working to provide direction and control for all the functions of the body). So while in common sense experience I believe that I am the sole agent of my own decisions, completely in charge of my own direction in life, in point of fact I am from moment to moment the end product of multiple interdependent physical systems at work within me. My mind, for example, in making a rational decision is heavily conditioned by the current state of my body (e.g., whether alert or fatigued), and my bodily organs work as well as they should only if my mind makes sensible rather than foolish decisions in its expenditure of time and energy. As Whitehead comments in *Process and Reality*, this understanding of a human being is quite different from the Aristotelian-Thomistic understanding of human nature as composed of body and soul, where the soul is the exclusive agent at work in the body and the body is the passive instrument of the soul in what it decides to be and to do.[36] For Whitehead, on the contrary, a human being is the moment-by-moment byproduct or result of the interrelated activity of all the agencies of the body located in the body's constituent actual entities making up its various subsocieties and nexuses.

The only place where Whitehead is not as clear as he should be is in his description of societies of actual entities as more than aggregates of individual entities. What he means by "more than an aggregate" is not apparent. A society is like a substance in classical metaphysics in that it endures over time with a specific pattern of existence and activity. But it is not like a substance in that its basic pattern of existence and activity undergoes change, is transformed, as a result of changes in its environment together with the ongoing response of its constituent actual entities to that environmental change. A Whiteheadian society is thus an evolutionary reality; a classical substance is a relatively fixed reality, resistant to anything but minor accidental change. This is why I have chosen to describe Whiteheadian societies as structured fields of activity for their constituent actual entities from moment to moment. A structured field of activity is an objective reality like an Aristotelian substance and thus more than an aggregate of subjective parts or components from moment to moment. It has an enduring structure or relatively fixed pattern of existence from moment to moment. But that structure never remains exactly the same; it is subject to gradual evolution in its mode of operation over time. In many ways, this field-oriented understanding of a Whiteheadian society mediates between the emphasis that Lazlo puts on systems as objective realities and the insistence of Whitehead

36. Ibid., 108–9.

that societies do not exercise agency in their own right but exist as byproducts or results of the collective agency of their constituent actual entities. Better than Whitehead, I can affirm with Laszlo that systems/societies can describe the objective hierarchical structure of nature from atoms to cells, organisms, communities, and environments on this earth, and beyond this earth to the clustering of stars into galaxies and sets of galaxies.[37] But with Whitehead I can argue against Laszlo that societies/systems do not possess agency in and of themselves but only in and through their component parts or members as dynamically interrelated subjects of experience.

To sum up, then, in this chapter I have briefly sketched how philosophers in the Western tradition have tried to achieve a sense of truth and objectivity for their speculative schemes. Plato and Aristotle had great confidence in the power of transcendental Forms or Ideas to illuminate sense data. Aquinas located these Forms in the mind of the biblical God as Creator of heaven and earth. Descartes turned to the human mind as the empirical focus for the data of experience and the ideas that it generates. Kant eventually proposed that the a priori structures of the human mind working on sense data are the source of truth and objectivity in human experience of the external world. The German Idealists constructed elaborate metaphysical systems to bridge the resulting gap between appearance and reality in human experience, but the purely conceptual or essentially closed character of their systems led to a widespread rejection in Western academic circles of the possibility of a meta-narrative or overarching metaphysical system. Yet the persistent desire both of scientists and of scholars in the humanities for logical consistency and internal coherence of ideas with one another in their research and publications has almost inevitably led to a revival of systems-oriented thinking in recent years. But this revived use of systems-oriented thinking carries with it the proviso that the system in question must be open-ended, capable of gradual evolution or perhaps even dramatic transformation, given the contingencies of existence in this world. In the next three chapters, I will pursue this concept of an open-ended system through comparison of the notion of a Whiteheadian society as a structured field of activity for its constituent actual entities with a systems-oriented approach to reality in three different areas: the natural sciences, the social sciences, and the workings of the democratic process in contemporary political life.

---

37. See Ervin Laszlo, *The Systems View of the World: The Natural Philosophy of the New Development in the Sciences* (New York: Braziller, 1972), 30–33.

6

Whiteheadian Societies as Self-Unifying Systems

In the introduction to his book *Darwin's Cathedral*, David Sloan Wilson asks, "What is the nature of human society? Is it a collection of self-seeking individuals, or can it be regarded as an organism in its own right?"[1] Noting that the understanding of groups as akin to organisms has in recent years been called into question by evolutionary biologists such as Richard Dawkins,[2] Wilson claims that recent advances in evolutionary biology using a methodology called "multilevel selection theory" can with proper qualifications give new life to the older commonsense belief that groups are equivalently organisms both in their internal organization and in their external behavior toward other groups.[3] But he cautions against thinking of all social groups as organisms, at least in light of the Darwinian principle of natural selection with its emphasis on survival and reproduction. That is, even though Darwin himself in *The Descent of Man* claimed that "the three ingredients of natural selection—phenotypic variation, heritability, and fitness consequences—can function at the level of groups,"[4] it does not follow that group members will invariably behave unselfishly toward one another and members of other groups. Some otherwise well-organized groups will have a majority of members who pursue largely self-centered ends and values and thereby unconsciously put at risk the ongoing survival of the group.

---

1. David Sloan Wilson, *Darwin's Cathedral: Evolution, Religion, and the Nature of Society* (Chicago: University of Chicago Press, 2003), 2.
2. See, for example, Richard Dawkins, *The Selfish Gene* (Oxford: Oxford University Press, 1976).
3. Wilson, *Darwin's Cathedral*, 17.
4. Ibid., 9. See also Charles Darwin, *The Descent of Man and Selection in Relation to Sex* (New York: Appleton, 1871), 166.

In any case, contrary to the views of evolutionary biologists like Richard Dawkins, not just genetics but likewise moral values play a key role in the evolution of religion and of other human social systems: "The behaviors that count as right conduct are not genetically determined but depend on open-ended psychological and cultural processes."[5] Hence, insofar as religion sets forth rules of right conduct for its adherents, it is a positive rather than strictly negative factor in the origin and survival of nonreligious social systems. "Otherworldly" religious beliefs spurned by atheists and other skeptics of institutional religion actually involve evolutionary advantages for those who embrace them. For they motivate altruistic behaviors that are adaptive in the real world. Hence, if one has to choose between "factual realism based on literal correspondence [with empirical reality] and a practical realism based on behavioral adaptedness,"[6] it makes more sense to be a practical realist. Rationality, after all, is ultimately subordinate to adaptation for the purposes of survival and reproduction. Science, to be sure, is based on a commitment to factual realism. But this could be seen as an argument for the complementarity of religion and science in the contemporary world. That is, the values proper to scientific research do not suffice for society as a whole. They must be supplemented with other values that place a greater emphasis on practical realism and that hopefully apply to all members of the society as moral equals.

My purpose in this chapter, however, is not simply to praise Wilson's perceptiveness in seeing the complementary workings of religion and science in contemporary Western society, but also to suggest that his hypothesis of open-ended "unifying systems" operative in nature through the evolutionary mechanism of group selection could well be confirmed and philosophically extended in terms of my revision of Alfred North Whitehead's notion of "society" as a collection of actual entities (momentary self-constituting subjects of experience) linked by a "common element of form."[7] Wilson, for example, claims that unifying systems exist throughout nature: "The same theory that explains human groups as adaptive units also explains social insect colonies, individual organisms, and even the origin of life itself as unified groups of interacting molecules that evolved by group selection."[8] Could one also apply the idea of unifying systems

---

5. Ibid., 223–24.
6. Wilson, *Darwin's Cathedral*, 228.
7. Alfred North Whitehead, *Process and Reality: An Essay in Cosmology*, corrected edition, ed. David Ray Griffin and Donald W. Sherburne (New York; Free Press, 1978), 34.
8. Wilson, *Darwin's Cathedral*, 222–23.

with internal self-organization to the inanimate world of atoms and even subatomic particles like electrons and neutrons on the assumption that they too are the ongoing result of a process of internal self-organization? Given that "actual entities" as "the final real things of which the world is made up"[9] are active at even the inanimate level of existence and activity within nature to produce societies, products of internal self-organization, why not make that further claim? But then one has to establish that Wilson's unifying systems and Whiteheadian societies are functionally equivalent. My argument in this chapter is that this is only possible if one accepts that Whiteheadian societies are more than groups of actual entities with a common element of form but are instead objective realities in their own right as structured fields of activity for their constituent actual entities from moment to moment.

Objections to my proposal will, of course, be immediately offered both by evolutionary scientists and by many more orthodox Whiteheadians. Evolutionary scientists might well balk at the idea of a psychic dimension to physical reality at the purely inanimate level of existence and activity within nature. Many Whiteheadians could object that in Whitehead's metaphysics societies are not objectively existing social realities ontologically distinct from their constituent parts or members. Societies are from moment to moment a collection of genetically related subjects of experience that share an affinity with one another in terms of a common element of form but still in their brief moment of existence are preoccupied with their own individual self-constitution rather than with their simultaneous cocreation of a social reality greater than themselves. Yet if I am right in my claim that Wilson's notion of unifying systems can be confirmed and further extended as a result of this revised understanding of Whiteheadian societies, then it should make a significant difference both to natural scientists and to Whiteheadians. For on the one hand, Wilson and other natural scientists would have a philosophical as well as a purely functional argument for the reality of adaptive unities or unifying systems operative throughout nature. That is, they could then claim that unifying systems are not simply conceptual tools for scientific research. Rather, they correspond to what is already going on in nature on a normal basis. For Whiteheadians, on the other hand, there is a distinct advantage in showing the relevance of Whitehead's metaphysical scheme for current empirical research in the natural sciences.

---

9. Whitehead, *Process and Reality*, 18.

Adaptive Unities and Whiteheadian Societies

In what follows, I first lay out what I see as notable similarities between Wilson's understanding of a unifying system and Whitehead's own definition of a society. Then I will explain how my revision of the notion of society in Whitehead's metaphysics aligns it much more closely with what Wilson proposes for unifying systems. To begin, Wilson claims that groups can be defined as a set of individuals who share a single trait.[10] Some individuals within the group may well share multiple traits, but what specifically defines the group as one group rather than another is a single trait or common activity:

> My bowling group is the people with whom I bowl, my study group is the people with whom I study, my platoon is the group of people with whom I fight, my nation is the people who share the same set of laws, my church is the group of people with whom I worship. All of these groups are defined in terms of the individuals who interact with respect to a given activity. There is an infinite variety of groups, but only when we consider an infinite variety of activities. For any particular activity, there is a single appropriate grouping.[11]

From Wilson's perspective, the reason why groups have to be thus identified with a single trait or common activity is that this is the only way to determine whether the trait or activity really defines the group as a group and not just as a set of individuals with roughly similar behavior patterns. For if one simply compares the average number of individuals who, while possessing the trait, survive and reproduce with the number of those who do not survive and reproduce in spite of possessing the trait, one proves nothing about the way a group as such has managed to survive. It would simply be a matter of natural selection operating among individuals in a given locality so that some survive and others do not.

Looking now at Whitehead's definition of a society in *Process and Reality*, we read the following:

> The common element of form is simply a complex eternal object [pattern or structure] exemplified in each member of the nexus [society]. But the social order of the nexus is not the mere fact of this common form exhibited by all its members. The reproduction of the common form throughout the nexus is due to the genetic relations of the members of the nexus among

---

10. Wilson, *Darwin's Cathedral*, 15–16.
11. Ibid.

each other, and to the additional fact that genetic relations include feelings of the common form. Thus the defining characteristic is inherited throughout the nexus, each member deriving it from those other members of the nexus which are antecedent to its own concrescence [self-constitution].[12]

In more commonsense language, what Whitehead is claiming is that the constituent members of the group feel their affinity with one another in terms of a feeling-level grasp of a common trait that all of them acknowledge as specific to themselves as this group rather than some other group. That trait may change character somewhat over time as new members are added and some currently existing members cease to participate in the group. But the group as a whole still has a "defining characteristic" that clearly marks it out as this group rather than another group.

Thus, Whitehead's concept of a society seems to correspond to Wilson's claim that every group should be defined by the possession of a common trait or pattern of existence and activity. Furthermore, if, as Whitehead says, actual entities are the final real things of which the world is made up, and if actual entities instinctively aggregate into societies at all levels of existence and activity within nature, then unifying systems are, as Wilson also proposes, operative everywhere that we find "things" that endure. Organisms and possibly even inanimate things are in the end groupings of components that work together in harmonious fashion. As Wilson comments, upon closer inspection organisms turn out to be social groups, a set of individual components or members, all of whom contribute in their own way to a higher-order unity and value.[13] Whitehead, for his part, contends that actual entities "prehend" (grasp on a feeling level) their affinity with and for one another in terms of a common element of form. Hence, Whitehead provides a philosophical explanation for the way in which Wilson's unifying systems come into existence and survive over time. All the components of the unifying system somehow recognize the single trait that marks them out as this group rather than another. They are not a group simply because of geographical proximity to one another or because of other contingent factors. They constitute a group because in some way they "feel" an affinity for one another and "want" to sustain it.

This does not necessarily presuppose, of course, intentionality or consciousness on the part of the components. For as noted above, Wilson thinks that unifying systems may well exist even at the level of interacting molecules in the origin of life in this world. What Whitehead proposes over

12. Whitehead, *Process and Reality*, 34.
13. Wilson, *Darwin's Cathedral*, 18.

and above what Wilson claims is that the interaction of molecules with one another is ultimately due to the fact that molecules are complex "structured societies" of atomic subsocieties, which in the end are composed of momentary subjects of experience in dynamic interrelation. Only with this presupposition of components as dynamically interrelated subjects of experience can one logically explain how apparently inanimate things like molecules "feel" an attraction for one another that results in their becoming not simply a purely contingent aggregate of material entities but, in Wilson's terminology, a unifying system with a capacity for growth and development. Wilson is quite confident that the ongoing "adaptive unity" between components of a unifying system does not happen strictly by chance in virtue of random genetic mutations, as some evolutionary biologists might claim. Nor is it a purely mechanistic, completely predictable event, as other evolutionary biologists might claim. It only happens on a trial-and-error basis with some unifying systems surviving long enough to reproduce themselves and others not. As he notes, "selfish" genes that gain at the expense of other genes are in the end detrimental to the survival of the organism as a unified whole. Yet given enough "unselfish" or "altruistic" genes working together within a unifying system, the organism prospers and succeeds in reproducing itself, thereby being one of the "survivors" in the cosmic process of natural selection.[14]

Yet this trial-and-error mode of operation for unifying systems raises the further metaphysical question of whether the workings of natural selection within unifying systems can and should be seen as an instance of Whiteheadian "creativity," an inbuilt spontaneity of nature at all levels of existence and activity. That is, in Whitehead's scheme, creativity empowers actual entities as self-constituting subjects of experience to make themselves to be what they are in virtue of their individual appropriation of the common element of form or defining characteristic of the society to which they belong. This rules out, of course, any form of strict determinism within Whiteheadian societies. At the same time, this self-constitution of any given actual entity is not simply a matter of chance. The actual entities within a society must in some measure conform to the pattern of coexistence and activity that their predecessor actual entities within the society have already established by their dynamic interrelation. Yet as Whitehead also insists, each actual entity has a unique self-identity; it is never fully identical with its contemporaries in the society to which it belongs.[15] So Wilson's presupposition of a trial-and-error approach in the origin and

---

14. Ibid.
15. Whitehead, *Process and Reality*, 22–23.

growth of unifying systems in nature certainly makes sense in light of the way creativity works within Whitehead's metaphysical scheme. There is no strict guarantee in advance how the actual entities constitutive of a given society will respond to one another from moment to moment. Only afterward, in the light of the persistent pattern of "decisions" made by these actual entities over a period of time, does natural selection, as Wilson claims, come into play to determine which of these societies will survive as an enduring unifying system and which will shortly be eliminated in virtue of their negative or "self-centered" tendencies in dealing with one another.

Prior to the publication of *Darwin's Cathedral*, Wilson coauthored with a well-known philosopher of science, Elliott Sober, a book titled *Unto Others* on the evolution and psychology of unselfish behavior.[16] Therein Wilson and Sober argued that, while altruistic behavior among human beings has generally been analyzed and explained in terms of motives on the psychological level of human existence, it also admits of an evolutionary biological explanation. The example used by Wilson and Sober is the desire to avoid pain, which can be explained from an evolutionary as well as from a psychological perspective: "There is an evolutionary reason that we have the desire to avoid pain. Pain is typically associated with bodily injury. Present-day organisms seek to avoid pain because this strategy was favored by natural selection. Wanting to avoid pain is psychologically ultimate but evolutionarily instrumental."[17] At the same time, altruism can be and often is an implicit factor for human beings in making difficult choices if it is seen as somehow contributing to the evolutionary goal of survival and reproduction for the group to which one belongs, even if it does not always do so for the individual himself or herself. "People *want* to experience pleasure and to avoid feeling pain. However, human beings have supplemented these hedonistic desires in two ways. What we want for ourselves extends beyond the desire for pleasant states of consciousness. And human beings, we believe, have ultimate desires concerning the welfare of others."[18]

Like Wilson and Sober, Whitehead too believes that there is a long evolutionary history behind the normal functioning of human consciousness. The traces of this evolutionary history are to be found in the workings of the unconscious in the human mind. Thus, in Whitehead's view, careful analysis of successive moments of consciousness yields insights into

---

16. Cf. Elliott Sober and David Sloan Wilson, *Unto Others: The Evolution and Psychology of Unselfish Behavior* (Cambridge, MA: Harvard University Press, 1998).
17. Ibid., 201.
18. Ibid., 327.

largely preconscious modes of operation and psychic structures that are likewise to be found among actual entities at all the other levels of existence and activity within nature. Yet there are also significant differences in the way that Wilson and Sober deal with feelings and desires as factors in human decision making and the way that Whitehead incorporates feeling and desire into the process of self-constitution for actual entities at nonconscious levels of existence and activity within nature. This difference in presuppositions comes especially to the fore in the different ways that Wilson and Sober, on the one hand, and Whitehead, on the other hand, deal with the notion of "proposition."

Wilson and Sober claim that for humans, believing and desiring are "propositional attitudes."[19] If one wants a drink of water, one has to represent to oneself (a) that there is water in a glass in front of oneself and (b) decide to drink it. If the proposition "There is water in the glass" turns out to be false because some other liquid is in the glass, then one has to decide whether to satisfy one's thirst by drinking this other liquid or by looking around for the availability of drinking water somewhere else. Other animal species, say Wilson and Sober, likewise entertain beliefs and desires that motivate them to action, but unlike human beings they do not entertain these beliefs and desires by mental representations or concepts of what they believe to be the case and how they plan to satisfy their desire as a result.[20] Yet their behavior is still governed by some combination of belief and desire that involves a sensual representation of what they want and how to get it.[21]

Whitehead also uses the term "proposition" in his analysis of the self-constitution of an actual entity. Every actual entity, even those constituting inanimate things, has both a physical pole and a mental pole. The physical pole prehends (grasps on a feeling level) raw sense data from its predecessors in the same society and from the outer world around it. The mental pole prehends an objective pattern in the sense data thus received. In the case of actual entities composing inanimate things, plants, and lower-level animal organisms lacking a central nervous system and brain, the actual entity usually repeats in its own self-constitution the pattern already present in the sense data. This is why actual entities that are components of inanimate things show no visible signs of life; there is usually no trace of novelty or spontaneity in their self-constitution from moment to moment. "Living" actual occasions with at least some degree of spontaneity or creativity, however, will sometimes link a new conceptual pattern, what

19. Ibid., 208.
20. Ibid., 209–10.
21. Ibid., 212–13.

Whitehead calls a propositional feeling, with the sense data and thereby bring into existence a new common element of form both for themselves and for their successors in the same society, indeed, for the world at large insofar as it can be affected in some minimal way by this change of pattern.

For Wilson and Sober, then, a proposition is a mental representation of something actually existing—for example, a glass with (presumably) water in it, together with the desire to satisfy one's thirst by drinking it. For Whitehead, a "proposition" is a statement of possibility, not actuality—namely, what could be rather than what is. A self-constituting actual entity must then "decide" whether or not to make this possible pattern of existence its own or not. Where Whitehead basically agrees with Wilson and Sober is in the belief that propositions inevitably lead to decision and action. A proposition is grounded in feelings and desires. It is not just an impersonal mental representation but "a lure for feeling, thereby providing immediacy of enjoyment and purpose."[22] Yet while Wilson and Sober are primarily interested in the linkage of belief and desire within human beings and other higher-order animals at an explicitly cognitional level, Whitehead is interested in the workings of belief and desire within actual entities at many different levels of existence and activity within nature. Thus, Whitehead treats propositions not so much as facts to be grasped by human beings and other higher-order animals with a central nervous system and brain, but as feelings that can be received and transmitted by all "living" actual entities, even those at work in animal organisms without a central nervous system and brain and in actual entities that are the components of plants.

This may seem to be a minor point of disagreement, but in my judgment it also presents an insight into the overall worldview held by the parties in question. For Whitehead as a metaphysician/cosmologist, the world is held together in the first place by the communication of feelings among a very broad range of momentary subjects of experience. For Wilson and Sober, on the contrary, with their specific interest in the workings of beliefs and desires on the psychological level for humans and higher-order animals, cognition in some form or other must likewise be present so as to account for the workings of those same beliefs and desires. Both standpoints, of course, are correct, but Whitehead's is clearly broader in scope.

Rethinking Whitehead's Notion of Society

But does this comparison between David Sloan Wilson's "unifying systems" approach to group survival within an evolutionary context and

---

22. Ibid.

Whitehead's notion of society as the necessary byproduct of the dynamic interrelation of actual entities from moment to moment hold up under more careful scrutiny? Whitehead himself, after all, claims that actual entities are the final real things of which the world is made up.[23] Hence, only actual entities and not the societies into which they aggregate possess an internal unity and exercise agency, the power of self-constitution. So in contrast to Wilson's approach to unifying systems as higher-order realities existing in their own right and exercising some measure of control over their component parts or members, Whitehead considers societies to be strictly derivative realities with no internal unity and agency proper to themselves. They are simply the collective effect or ongoing result of the de facto interrelatedness of their component actual entities to one another. The ontological status of societies, however, would be dramatically different if Whitehead had extended the unifying power of creativity to the construction of societies as well as to the self-constitution of actual entities. That is, if creativity is the activity whereby "the many become one and are increased by one,"[24] then societies too could be seen as instances of the many becoming one and being increased by one. For societies, like actual entities, are instances of unity in diversity of parts or members. Likewise, they too, like actual entities, are "increased by one" in that with new component actual entities at every moment, a society is also numerically a new reality at every moment, albeit one with much the same internal structure or pattern of intelligibility as in the preceding moment. Most importantly, however, if creativity is at work in the constitution of societies, then they too, like actual entities, possess an internal unity, as opposed to the external unity of an aggregate. They thereby become entities in their own right over and above the activity of their component actual entities. Actual entities would still be necessary for a society to exist. But once originated, a society can be said to exist as a specifically social, objective reality (much akin to Wilson's "adaptive unities" or "unifying systems").

Why then did Whitehead refrain from extending this key metaphysical category of creativity to societies? His earlier-cited description of a society as a group of actual entities genetically related to one another seems to make clear that for him a society is much more than a contingent aggregate of similarly self-constituted actual entities. Yet by restricting the operation of creativity simply to actual entities and by not seeing creativity as likewise operative in the original formation and ongoing preservation of societies, he was logically committed to metaphysical atomism. As he

---

23. Whitehead, *Process and Reality*, 18.
24. Ibid., 21.

himself says in *Process and Reality*, "the ultimate metaphysical truth is atomism. The creatures are atomic."[25] Thus, in one way or another, societies are for him strictly derivative realities with no ontological status as specifically social realities distinct from the ongoing interplay of their constituent actual entities.

Here Whitehead may have been insufficiently critical of the strictly analytical approach to physical reality characteristic of early modern natural science up to his own day. That is, even though he rejected the implicit materialism of a mechanistic worldview in which atoms are simply inert material entities moved around by purely external forces and thus totally lacking in spontaneity, he did not further question whether actual entities, when understood as "spiritual atoms" in various combinations, fully account for the existence of different levels of existence ontologically distinct from one another within the natural world. Even more likely, he perhaps feared that in giving societies an ontological reality distinct from their component actual entities, he would be in danger of relapsing into the substance-and-accident metaphysics of Aristotle and the medieval scholastics. In *Adventures of Ideas*, for example, he admits that, given their expected endurance over time as opposed to the moment-by-moment existence of actual entities, societies are in some ways comparable to Aristotelian substances.[26] But the comparison ends there, and no explanation is offered as to how they differ. In any event, in what follows I summarize my alternative explanation of a Whiteheadian society as a structured field of activity for its constituent actual entities from moment to moment. For if I am right, this revision of the category of society brings Whitehead's process-oriented metaphysics into close alignment with Wilson's understanding of unifying systems within nature.

In brief, if a Whiteheadian society is understood to be an enduring structured field of activity for its constituent actual entities, it is neither simply an aggregate of similarly constituted actual entities nor an Aristotelian substance. That is, in opposition to the notion of an aggregate, a Whiteheadian society has an internal principle of order and intelligibility. It is not simply the contingent result of the ongoing interaction of its components. In this respect it resembles an Aristotelian substance. But in contrast to the classical understanding of substance, that internal principle of order and intelligibility is itself subject to gradual change or evolution with the passage of time, given that its constituents, momentary self-constituting subjects of experience, prehend not only one another but

---

25. Ibid., 35.
26. Whitehead, *Adventures of Ideas* (New York: Free Press, 1967), 204.

also changes in their external environment that somehow impact their own existence and activity. Much better than an Aristotelian substance, therefore, a Whitehead society as an enduring structured field of activity for its constituent actual entities is able to offer a philosophical explanation for the fact of ongoing evolution within the cosmic process. Change takes place in an orderly manner, but things do not remain the same over an extended period of time.

Yet one may still question whether a Whiteheadian society is capable of unified action as well as unified existence. An Aristotelian substance, for example, as a composite of immaterial form and material components is said to act in virtue of its immaterial substantial form. What is the corresponding principle of activity within a structured field of activity or, as Wilson would say, a unified system? Here Whitehead's notion of structured societies, that is, societies composed of subsocieties of actual entities,[27] provides an answer, albeit with some further qualification on my part. Whitehead claims that a structured society is actually composed not simply of subordinate societies but of subordinate nexuses as well.[28] Subsocieties of actual entities enjoy a certain independence of the structured society within which they here and now exist in that they could exist on their own apart from the structured society. Whitehead's example is that of a molecule within a living cell; it would still be a molecule of a certain type even apart from participation in the life of the cell. Subordinate nexuses, on the contrary, cannot maintain themselves in existence apart from participation in the life of the structured society as a whole. Such would be a nexus of "entirely living" actual entities with considerable novelty in their self-constitution from moment to moment but for the same reason with an inability to sustain a common element of form among themselves apart from the support provided by the subsocieties of nonliving actual entities coexistent with it in the structured society as a whole. What Whitehead clearly has in mind here is what Aristotle called the "soul" as distinct from the "body." The soul is the life principle of the body, its internal principle of existence and activity. Yet for the same reason (at least for Aristotle, if not for the medieval scholastics), the immaterial soul cannot exist apart from the body as the necessary physical infrastructure of its own more spiritual existence and activity.

Whitehead would side with Aristotle on this last point and even go beyond Aristotle in the claim that the nexus of entirely living actual entities within a structured society that corresponds to the classical notion

---

27. Whitehead, *Process and Reality*, 99.
28. Ibid.

of the soul is not different in kind, but only in degree of originality of self-expression, from the nonliving subsocieties of actual entities within the structured society as a whole.[29] Elsewhere in *Process and Reality* he comments, "an 'entirely living' nexus is, in respect to its life, not social. Each member of the nexus derives the necessities of its being from its prehension of its complex social environment; by itself the nexus lacks the genetic power which belongs to 'societies.'"[30] Yet despite the absence of a common element of form from moment to moment of its existence, an entirely living nexus "may support a thread of personal order along some historical route of its members."[31] That is, there is continuity between successive actual entities within the entirely living nexus, but the potentiality of significant change from moment to moment within an entirely living nexus is much higher than in the case of subsocieties of nonliving actual entities. Yet such nexuses of entirely living actual entities are not different in kind but only in the degree or "grade" of complexity from their nonliving counterparts within the structured society as a whole.

What I would add to Whitehead's account here of an entirely living nexus of actual entities within a structured society is that this entirely living nexus of actual entities is still just one subsociety within the structured society as a whole. What acts and thereby makes an impact on its environment is in the end not the nexus of entirely living actual entities on its own but the structured society as a whole functioning in and through all its subsocieties and subordinate nexus(es), even if principally through its dominant nexus of entirely living actual entities or "soul." This is not simply a small detail of Whiteheadian scholarship. At stake here once again, of course, is the question of the difference between a society and an aggregate of actual entities. Charles Hartshorne in an essay titled "The Compound Individual" claimed, on the one hand, that an aggregate of actual entities composing a rock or some other inanimate thing should be called a *composite* individual because it is no more than the sum of its parts but, on the other hand, that a Whiteheadian structured society with a "regnant" nexus of entirely living actual entities should be considered a *compound* individual because it has an internal unity proper to itself over and above the interplay of its constituent parts.[32] Yet this distinction seems

---

29. Ibid., 177–78, where Whitehead distinguishes between different "grades" of actual entities.
30. Ibid., 107.
31. Ibid.
32. Charles Hartshorne, "The Compound Individual," in *Philosophical Essays for Alfred North Whitehead*, ed. F. S. C. Northrup (New York: Russell & Russell, 1936), 215–17.

to lead to the affirmation of something like a "soul" within molecules and primitive forms of cellular life.[33] Far better, it seems to me, is to propose that every Whiteheadian society without exception is from moment to moment the product of all the various individual agencies at work within it, even though some of these agencies may play a greater role in the final outcome than others.

What is ultimately at stake here are two different models for the proper relation between the One and the Many. The older model, implicitly endorsed by Hartshorne and dating back to the philosophy of Plato, stipulates that the Many are ordered to one another in and through their common relationship to the One as a higher-order entity. The later model, which I endorse instead, claims that the Many by their dynamic interrelation from moment to moment produce the reality of the One as the structured field of activity for their conjoint activity.[34] Not a higher-order entity, therefore, but a common ground or structured environment for the interaction of the Many with one another constitutes the reality of the One within this later model. As I see it, this later model for the relationship between the One and the Many works much better than the older model in justifying an evolutionary approach to reality because it allows for incremental change in the nature or essence of the One that reflects the ever-changing reality of the Many in their interaction with one another and with the external environment. Using the older model, on the contrary, one is logically constrained to say that the One (like a substantial form in the philosophy of Aristotle and Thomas Aquinas) never changes in its basic structure and intelligibility. Change only takes place at the level of the Many insofar as they embody the enduring reality of the One in different ways. In this respect, Hartshorne was unconsciously reverting to a substance-oriented metaphysics in trying to define the difference between a compound individual and a composite individual in Whitehead's metaphysical scheme.

To sum up, then, a Whiteheadian society as a structured field of activity for its constituent actual entities (or a unifying system / adaptive unity in Wilson's scheme) not only exists as a unitary reality but likewise acts as a unitary reality or quasi-organism in its dealings with the world around it. But here one must be careful likewise to state that this agency of the

---

33. See David Ray Griffin, "Of Minds and Molecules," in *The Reenchantment of Science: Postmodern Proposals* (Albany, NY: State University of New York Press, 1988).

34. As already noted, I have studied this issue of two different models for the relationship between the One and the Many both historically and systematically in *Subjectivity, Objectivity, and Intersubjectivity: A New Paradigm for Religion and Science* (Philadelphia, PA: Templeton Press, 2009).

society or system is a collective agency, derivative from all the individual agencies at work within it. There is no single higher-order agency that unilaterally controls all the other lower-order agencies within the society or adaptive unity. All are needed, albeit in different degrees, for the society or adaptive unity to function as a unitary reality or quasi-organism. As a human being, I am not in the first place a mind or soul that uses its body to execute its internal desires. I am a body-soul unity that uses all the individual agencies of mind and body within me to make my presence felt in the world around me. In this sense, I am from moment to moment what Wilson calls a "unifying system" and what Whitehead terms a complex "structured society" of subsocieties of actual entities, all of which contribute to my physical and mental well-being from moment to moment. Likewise, I am an open-ended system in that whether I ultimately survive the weeding-out process of natural selection is contingent partly upon the decisions that I make from moment to moment and partly upon environmental and societal factors beyond my control. My future existence and activity, accordingly, is not totally "up for grabs" at every moment, but it is likewise certainly not predetermined in every detail.

7

Subjectivity and Objectivity within Open-Ended Systems

Many contemporary philosophers and theologians have a deep distrust of classical metaphysics with its strong focus on logical analysis and organized thought systems purporting to give a comprehensive view of the God-world relationship or some other all-embracing topic. This attitude may partly be traced to the impact on the academy of *Totality and Infinity* by Emmanuel Levinas, in which he laid bare the contrast between "totalizing" rational modes of thought and the potential infinity of human subjectivity as seen, above all, in the "face" of the other.[1] Likewise, the work of Jacques Derrida, Michel Foucault, and others in "deconstructing" classical texts in philosophy and theology so as to reveal their hidden mechanisms for power and control have made the rest of us alert to the subtle dangers of allegedly objective modes of thought. Yet reliance on subjectivity or a purely first-person perspective does not carry much weight in scientific research, where formal logic and systematic thinking are more or less taken for granted. So there is a tension between particularity, the inevitably subjective character of all human cognition, and the classical ideal of universal objectivity. In what follows I will first indicate how Whitehead and some of his followers (including Reiner Wiehl) may have overemphasized the role of subjectivity in nature through their almost exclusive focus on actual entities as "the final real things of which the world is made up."[2] Then I will sketch the approach to systems theory of

---

1. Emmanuel Levinas, *Totality and Infinity: An Essay on Exteriority*, trans. Alphonso Lingis (Pittsburgh, PA: Duquesne University Press, 1969), 21–30.
2. Alfred North Whitehead, *Process and Reality: An Essay in Cosmology*, corrected edition, ed. David Ray Griffin and Donald W. Sherburne (New York: Free Press, 1978), 18.

the noted sociologist Niklas Luhmann, who seems to have downplayed or even eliminated altogether the role of subjectivity in his strictly objective, purely functional analysis of how nature works. Finally, I will indicate how my own reinterpretation of Whiteheadian societies as structured fields of activity for their constituent actual entities manages to give equal value to the necessary interplay of subjectivity and objectivity both in systems thinking and in the workings of nature.

The Priority of Subjectivity over Objectivity

The late Reiner Wiehl worked hard to reconcile the demands of rational objectivity and creative spontaneity within human subjectivity in a series of books and articles over the years. In the year 2000, for example, he published *Subjectivität und System*, in which he pointed to the different ways in which the notion of subjectivity can be employed first in the construction of objective rational systems and then in the interpretation of subjective worlds of experience.[3] For my purposes in this chapter, however, I will focus on his essay "Whitehead's Cosmology of Feeling between Ontology and Anthropology," in *Whitehead's Metaphysics of Creativity*, edited by Wiehl himself and Friedrich Rapp.[4] Therein he compares Whitehead and Hegel on the theoretical presuppositions of a philosophy of life. Both Whitehead and Hegel, for example, rejected the classical notion of substance as the foundational concept of metaphysics. For substances or things only exist in the human mind as an unconscious abstraction from the dynamic flow of events taking place in reality. Hence, to focus on presumed substances or things is to commit what Whitehead calls "the fallacy of misplaced concreteness," to mistake the abstract for the concrete.[5] Likewise, both thinkers turned to the notion of subjectivity as the true source for a genuine philosophy of life. But at this point they differed significantly in their metaphysical schemes. Hegel conceived his system as both the logical and historical self-manifestation of Absolute Spirit wherein objectivity and subjectivity are in the end perfectly reconciled. But Hegel in that respect was implicitly committed to a totalizing approach to reality. The process of the self-manifestation of Absolute Spirit

---

3. Reiner Wiehl, *Subjectivität und System* (Frankfurt am Main: Suhrkamp, 2000).
4. Reiner Wiehl, "Whitehead's Cosmology of Feeling between Ontology and Anthropology," in *Whitehead's Metaphysics of Creativity*, ed. Reiner Wiehl and Friedrich Rapp (Albany, NY: State University of New York Press, 1990), 127–51.
5. Alfred North Whitehead, *Science and the Modern World* (New York: Free Press, 1967), 51.

was already logically complete in his philosophy and in due time would presumably also be fully manifest in history. When the triumph of rationality in cosmic history is complete, then the process character of life will inevitably come to an end.

Whitehead, on the contrary, with his turn to the subject, conceived reality in terms of a quasi-infinite number of momentary self-constituting subjects of experience that by their dynamic interrelation from moment to moment coconstitute the apparent substances or enduring things of this world.[6] As Wiehl perceptively comments, Whitehead thereby privileged creativity or life over rationality or truth.[7] Given the never-ending multiplicity of centers of subjectivity, each engaged in its creative self-constitution out of the data presented to it by antecedent subjects of experience, there can be no end to the ongoing process of life. Truth is therefore never an accomplished fact, as in the speculative philosophy of Hegel: "The reality of life and of the individual processes of life is more original than the truth of knowledge and process of its proof of truth. Truth is only one of many functions of life, as incidentally also error and deception, although a very important one: It encourages the development of higher life."[8] As Whitehead himself commented in *Adventures of Ideas*, "it is more important that a proposition be interesting than that it be true. . . . But, of course, a true proposition is more apt to be interesting than a false one."[9] What is ultimately important is the ongoing process of life, even if it sometimes meanders into dead ends along the way.

Yet is it possible that Whitehead overplayed the priority of subjective creativity over objective rationality? Did he, in other words, adequately account for the existence of rationality and order within the creative process? Did he not postulate, for example, the necessity of the consequent nature of God in *Process and Reality* so as to bring into harmony what might otherwise be a chaotic jumble of disconnected events? "God and the World are the contrasted opposites in terms of which Creativity achieves its supreme task of transforming disjoined multiplicity, with its diversities in opposition, into concrescent unity, with its diversities in contrast."[10] Yet if so, is this not a deus ex machina for what should be an intrinsic feature of reality even apart from God? Whiteheadians might counterargue that in virtue of Whitehead's ontological principle the ultimate reasons for

---

6. Whitehead, *Process and Reality*, 18, 34–35.
7. Wiehl, "Whitehead's Cosmology of Feeling," 146–50.
8. Ibid., 147.
9. Alfred North Whitehead, *Adventures of Ideas* (New York: Free Press, 1967), 244.
10. Whitehead, *Process and Reality*, 348.

things are to be found in actual entities.[11] So what is the problem with stipulating that the ultimate unity of things in this world is to be found within God, the primordial actual entity? But this is clearly an instance of what postmodernists like Jacques Derrida have called "logocentrism," the appeal to a "transcendental signified" so as to satisfy the human passion for order and intelligibility.[12] The unity of things should rather be found in the things themselves and in their internal relations to one another, not in appeal to a "transcendental signified" like the divine consequent nature in Whitehead's philosophy.

Likewise, in Whitehead's definition of society in *Process and Reality*,[13] the "common element of form" for the society is said to arise in each member of the nexus by reason of its prehensions of that form in "some other members of the nexus." Hence, each constituent actual occasion has to filter out what is irrelevant as opposed to what is relevant for itself in its process of self-constitution. Furthermore, it does so by focusing its attention on only some of its antecedent actual occasions but, by implication, not on all of them. Hence, the question naturally arises as to how one can be sure that the same common element of form or governing pattern of dynamic interrelation will in fact be faithfully transmitted from one set of actual occasions to another so as to perpetuate the "defining characteristic" of the society as a whole.

Whitehead's answer presumably would be that it is not necessary for the common element of form to be reproduced in exactly the same way within each newly concrescing actual occasion of the society in question. As he notes elsewhere in *Process and Reality*, it is enough that the constituent actual occasions are analogously the same in terms of their common reproduction of the defining characteristic of the society.[14] But once again, the question arises: How one can be sure that the society will not cease to exist someday or at least not dramatically change character from one moment to the next? Each constituent actual occasions is, after all, a world unto itself. As Whitehead himself comments, "no two actual entities originate from an identical universe; though the difference between the two universes only consists in some actual entities, included in one and not in the other, and in the subordinate entities which each actual entity

---

11. Ibid., 19.
12. Jacques Derrida, *Of Grammatology*, trans. Gayatri Chakravorty Spivak (Baltimore, MD: Johns Hopkins University Press, 1976), 10–26.
13. Whitehead, *Process and Reality*, 34.
14. Ibid., 89.

introduces into the world."[15] Among these "subordinate entities" would be presumably eternal objects that "are the same for all actual entities"[16] but may not be prehended by each concrescing actual occasion in the same way. So there is an objective uncertainty about the continuity over time of Whiteheadian societies with a well-defined common element of form. Given the evolutionary character of physical reality according to Whitehead, one would expect that common element of form to change gradually over time. But what logical guarantee is there that it will not dramatically change from one moment to the next and effectively be a different society or no society at all?

Finally, by his own admission, Whitehead was a philosophical atomist: "The ultimate metaphysical truth is atomism. The creatures are atomic."[17] Michel Weber has pointed out, to be sure, that Whitehead's emphasis on actual entities as the final real things of which the world is made up was necessary to allow for genuine novelty in an evolutionary context.[18] Without the privacy afforded by the notion of a momentary self-constituting subject of experience that is heavily influenced but not controlled by its environment, there is no ontological basis for spontaneity in a cosmic process otherwise dominated by fixed patterns of activity. Yet did Whitehead thereby vindicate the reality of novelty in his evolutionary cosmology at too high a price? For if, as noted above, "no two actual entities originate from an identical universe," then there is no objectively existing universe at all. Even the universe as existing from moment to moment within the consequent nature of God is not *the* universe but *God's* universe, the panoply of events in this world as experienced by God and somehow integrated into a harmonious totality from moment to moment. The ontological unity thus achieved only exists within God as the primordial actual entity, not between finite actual entities in this world. Thus there is no objectively existing universe within Whitehead's metaphysical scheme because all ontological unity is the unity of actual entities in their individual self-constitution.[19] There are unquestionably close analogies between actual entities, on which basis Whitehead can stipulate the existence of societies with a common element of form among their members, but there is no

---

15. Ibid., 22–23.
16. Ibid., 23.
17. Ibid., 35.
18. Michel Weber, "Introduction: Process Metaphysics in Context," in *After Whitehead: Rescher on Process Metaphysics*, ed. Michel Weber (Heusenstamm bei Frankfurt: Ontos Verlag, 2004), 58–59, 64; see also Whitehead, *Process and Reality*, 18.
19. Whitehead, *Process and Reality*, 26.

universe as such. For in the end, as Whitehead himself maintains, actual entities are the final real things of which the world is made up.

Admittedly, at the very end of *Process and Reality* Whitehead sets forth the "four creative phases in which the universe accomplishes its actuality."[20] But careful analysis of those phases makes clear that the third phase, "the phase of perfected actuality, in which the many are one everlastingly, without the qualification of any loss either of individual identity or of completeness of unity,"[21] is a description of the divine consequent nature, not of the world, except as it is indirectly reflected in the former. Whitehead then adds in the fourth phase that "the perfected actuality passes back into the world, and qualifies this world so that each temporal actuality includes it as an immediate fact of relevant experience."[22] Through prehension of the divine consequent nature by each newly concrescing finite actual occasion and with the provision by God of an "initial aim" to initiate its process of self-constitution,[23] God's experience of the deeper harmony of events taking place in this world becomes a factor in that entity's self-constitution. But it is still only one factor among many others at work in the process of concrescence for that actual occasion. In the end, accordingly, we are left with a multiplicity of entities, each engaged in its own self-constitution. There is no objectively existing universe.

Was this a mental lapse on Whitehead's part or a deliberate choice? In all likelihood, it was a deliberate choice since, as already noted, he consciously espoused metaphysical atomism. But perhaps the deeper reason for that choice may have been that he had no other logical alternative beyond atomism, given his repudiation of the classical notion of substance. It is revealing that in *Adventures of Ideas* he aligns his own understanding of societies with the classical notion of substance, even as he notes that confusion between the two terms has been an obstacle in Western philosophy since the time of the Greeks:

> The real actual things that endure are all societies. They are not actual occasions. . . . A society has an essential character, whereby it is the society that it is, and it also has accidental qualities which vary as circumstances alter. Thus a society, as a complete existence and as retaining the same metaphysical status, enjoys a history expressing its changing reactions

---

20. Ibid., 350–51.
21. Ibid., 350.
22. Ibid., 351.
23. Ibid., 244.

to changing circumstances. But an actual occasion has no such history. It never changes. It only becomes and perishes.[24]

Whitehead evidently thinks that it is a mistake to confuse actual occasions with minisubstances despite the fact that he himself refers to actual occasions as the final real things of which the world is made up. Societies, on the contrary, are much closer in ontological function to substances in that they endure over time. But what are societies if they have enduring thing-like characteristics and yet seem to be only collections of analogously constituted actual occasions, as noted above? Here Whitehead may have been at a loss to say, and so he simply devoted his attention to the self-constitution of the constituents of societies, actual occasions, and their dynamic interrelation.[25]

While I agree with Reiner Wiehl, therefore, that Whitehead's metaphysical system is a better candidate for a comprehensive philosophy of life than that of Hegel because of its strong emphasis on creativity and ongoing process, even here rationality and the persistence of an established order over time have to be safeguarded. Aristotle's category of substance, in other words, cannot be simply set aside as no longer necessary in a process-oriented world. Its equivalent in terms of a principle of continuity of form or pattern of operation over time must be acknowledged and defined even within a process-relational worldview. To help make this point clear, I now cite the work of two prominent non-Whiteheadian but still process-oriented thinkers, Ivor Leclerc and Ervin Laszlo, who like me were dissatisfied with Whitehead's philosophical atomism and sought to remedy it in terms of a somewhat different metaphysical scheme. I will then indicate how my own revision of Whitehead's philosophy seems to be both more faithful to Whitehead's approach to reality and better suited to the requirements of a bona fide social ontology.

Ivor Leclerc in *The Philosophy of Nature* first indicates how Gottfried Leibniz and Whitehead sharply critiqued the materialistic atomism of

---

24. Whitehead, *Adventures of Ideas*, 204.
25. See here James W. Felt, *Coming to Be: Toward a Thomistic-Whiteheadian Metaphysics of Becoming* (Albany, NY: State University of New York Press, 2001), 43–80. Felt's key category is "primary being," that is, a subject of activity existing by participation in the ultimate source of existing and activity that is God (62–63). In this way, he retains Whitehead's focus on subjects of experience as "the final real things of which the world is made up" but provides for continuity within subjective experience by reverting to what he regards as the original Aristotelian understanding of substance. Felt and I, therefore, share the same misgivings about Whitehead's metaphysical atomism but seek to resolve it in different ways.

early modern Western philosophy but curiously retained some of its basic features, which then created problems for them in explaining "compound individuals," actualities composed of interrelated parts or members. Leibniz claimed that individual monads making up a physical body each perceive the body as a whole "in a particular perspectival relationship."[26] A substantial bond (*vinculum substantiale*) thereby exists between the monads making up the body. But this is ultimately an illusory claim, says Leclerc, since in the end only the monads really exist. Similarly, Whitehead claimed that, since actual occasions are the final real things of which the world is made up, societies of actual occasions are strictly derivative from the dynamic interrelatedness of those same actual occasions. Yet if the reasons for things are to be found only in actual entities,[27] a society as such has no ontological reality or reason to exist: "the society is not 'self-sustaining' except in a derivative sense; it is the constituent actual entities which individually 'sustain' the defining character of the society."[28]

Leclerc's own solution to the problem of compound individuals is to return to Aristotle's notion of a "mixture," whereby the constituent parts or members are integrated with one another so as to form a new substantial reality with its own form or agency. The constituent parts or members thereby become potentialities with respect to the new actuality to be found in the whole. Only if the compound individual dissolves do the component parts or members become once again fully actual entities in their own right. Leclerc summarizes his position as follows:

> There can be a unity of a compound only by the unifying of the constituents, by their integration into a new whole. Now this unifying is an act, and this implies an agent acting, i.e., an agent effecting this unity. This unifying agent cannot be itself one of the constituents unified, for it could not then be a constituent. If there is to be an integral unity qua compound, the unifying agent must transcend the constituents. This implies that the actuality of the unifying agent must itself be emergent in the unifying.[29]

---

26. Ivor Leclerc, *The Philosophy of Nature* (Washington, DC: Catholic University of America Press, 1986), 117. See also by the same author *The Nature of Physical Existence* (New York: Humanities Press, 1972), 289: "what exists in actuality is a mere aggregate, and the members of the group act in a certain harmony, which gives to it an appearance of a group character, by virtue of a co-ordination pre-established by God of the actions of the constituent monads which are in themselves separate from each other."

27. Whitehead, *Process and Reality*, 19.

28. Leclerc, *The Philosophy of Nature*, 120; *The Nature of Physical Existence*, 289–91.

29. Leclerc, *The Philosophy of Nature*, 127; *The Nature of Physical Existence*, 296.

My own position is quite close to that of Leclerc, but it is still different. As noted in earlier chapters, in my field-oriented understanding of Whiteheadian societies, the actual occasions making up a given society coconstitute from moment to moment their common element of form. But the common element of form thus constituted exercises in turn another kind of causality; that is, it "informs" or provides a governing structure for the next set of constituent actual occasions in terms of their individual processes of concrescence. Simply by being available for prehension and thereby for integration into their individual processes of concrescence, the common element of form exercises top-down causality upon the next set of actual entities within the society. In this way I remain faithful to Whitehead's dictum that agency in the sense of efficient causality "belongs exclusively to actual occasions."[30]

After all, Aristotle and Leclerc's notion of a compound individual works reasonably well to explain the emergence and functioning of complex individual organisms, but it does little or nothing to explain how supraorganic realities like communities and physical environments over time assume an "essential character" that is different from the combined characteristics of their individual parts or members. Here, as I see it, is where my own understanding of a Whiteheadian society as a structured field of activity for its constituent actual occasions has an applicability far beyond the Aristotelian category of substance. For if one conceives communities and environments as likewise substances or compound individuals, then totalitarianism in some form or other results. That is, the parts or members of the community or environment thereby lose their individual self-identity through being absorbed into the unitary actuality of the environment or community. The model of a physical organism thus does not work well for the analysis of communities and environments where the component parts or members must remain fully actual realities in their own right.

I turn now to the work of Ervin Laszlo, one of the early theoreticians of "systems theory." In his book *Introduction to Systems Philosophy*, Laszlo first distinguished between humanly constructed or artificial systems and natural systems. A natural system he then defined as a "non-random accumulation of matter-energy, in a region of space-time, which is nonrandomly organized into coacting interrelated subsystems or components."[31] As such, the notion of natural system applies not only to individual entities, both nonliving and living, but also to specifically social groupings, namely,

---

30. Whitehead, *Process and Reality*, 31.
31. Ervin Laszlo, *Introduction to Systems Philosophy: Toward a New Paradigm of Contemporary Thought* (London: Gordon and Breach, 1972), 30.

communities or environments. Furthermore, he sees the entire physical universe as thus hierarchically organized into ever-more complex and/or comprehensive systems.

Thus understood, a natural system for Laszlo is much akin to my own understanding of a Whiteheadian society as a structured field of activity for its constituent actual occasions, with two significant exceptions. For Laszlo, primitive natural systems like atoms possess "invariant properties" that allow them to become functioning parts of more sophisticated natural systems.[32] Following Whitehead, I instead hold that actual occasions as momentary self-constituting subjects of experience have no invariant properties but derive their internal structure and organization here and now from the environment out of which they are emerging. Likewise, with Whitehead I hold that Whiteheadian societies (unlike natural systems for Laszlo) do not exercise agency in their own right but only in and through the collective agency of their constituent actual occasions from moment to moment. I thereby locate agency or efficient causality in the constituents of a society or system rather than in the society or system itself, even though I agree with Laszlo that societies or systems are ontological realities in their own right.

My point here is not to vindicate my own understanding of a Whiteheadian society over against other alternatives (compound individual for Leclerc, natural system for Laszlo) but to set forth a foundational concept for a new social ontology, one that does not reduce the constituent parts or members to the whole or the whole to the interplay of the constituent parts or members. Laszlo's natural system evidently provides for the existence and activity of specifically social realities distinct from their constituent parts or members. But in attributing agency directly to the natural system instead of seeing the agency of the system as the collective agency of its interrelated parts or members, Laszlo runs the risk of totalitarianism. The natural system thereby subordinates its parts or members to its own superordinate agency, much like Leclerc's notion of the substantial form. The parts or members are then no longer actualities in their own right but instead potentialities within the actuality of the natural system. As a result, systems no longer exist for the sake of the individuals who created them; rather, individuals exist to perpetuate the goals and values of the system.[33]

---

32. Ervin Laszlo, *The Systems View of the World: The Natural Philosophy of the New Developments in the Sciences* (New York: Braziller, 1972), 30–33.

33. See Joseph A. Bracken, *The One in the Many: A Contemporary Reconstruction of the God-World Relationship* (Grand Rapids, MI; Eerdmans, 2001), 142.

To sum up this first part of the chapter, while I fully agree with Reiner Wiehl that within a comprehensive philosophy of life creativity and process should have priority over rationality and order, a suitable balance between these rival principles of existence and activity must be ultimately established. Whitehead, in my judgment, overplayed the role of creativity within his metaphysical system, even though, as Friedrich Rapp makes clear, he evidently wanted to achieve a systematic understanding of reality with a precision akin to that found in the natural sciences.[34] His mistake lay in focusing too much on the reality and function of actual entities as momentary self-constituting subjects of experience and too little on the reality and function of the societies out of which they emerge and to which they contribute their own feeling-laden pattern of objectification. Only if "actual entity" and "society" are seen as equiprimordial realities rather than as the latter ontologically derivative from the former is one in a position to strike the proper balance between subjective creativity and objective rationality, novelty and order. Not subjectivity by itself, therefore, but only intersubjectivity is the necessary starting point for a comprehensive philosophy of life, provided that by intersubjectivity one presupposes the notion of a Whiteheadian society as a field of activity progressively structured and gradually transformed by the ongoing interplay of its constituent actual entities.

The Priority of Objectivity over Subjectivity

In his groundbreaking work *Being and Time*, Martin Heidegger proclaimed the end of classical metaphysics. Since it was based on an unconscious confusion of Being in itself with God as the Supreme Being, classical metaphysics in Heidegger's view lacked real objectivity. For it never addressed the true reality of Being as that which manifests itself at intervals to *Dasein*, defined as a human being who reflects on the contingency ("thrownness") of her own existence and seeks to achieve intelligibility and order in her life through a self-constituting decision.[35] Because Heidegger was also critical of the classical notion of substance in terms of traditional subject-object or subject-predicate relations,[36] the influence of his thought

---

34. Frederich Rapp, "Whitehead's Concept of Creativity and Modern Science," in *Whitehead's Metaphysics of Creativity*, ed. Reiner Wiehl and Friedrich Rapp (Albany, NY: State University of New York Press, 1990), 70–93.

35. Martin Heidegger, *Being and Time*, trans. John Macquarrie and Edward Robinson (New York: Harper & Row, 1962), 67–77, 312–15.

36. Ibid., 71–75.

was clearly felt in still other antimetaphysical positions such as structuralism, poststructuralism, and deconstructionism. But the persistent need for some kind of objectivity in the natural and social sciences eventually led to the development of systems theory in the natural and social sciences. Systems theory focuses on the objective rule-governed context of observable events rather than on the human and nonhuman agents at work in those contexts. Human subjectivity and other forms of subjectivity within nature are, of course, thereby reduced to being no more than sine qua non conditions for the operation of an objective system.[37] In this sense, systems theory is postmetaphysical, at least in the mind of Niklas Luhmann, one of the principal proponents of systems theory in the late twentieth century. For it basically eliminates the need for metaphysics as the ultimate rational explanation of the way things work within this world.[38]

Even within systems theory, however, interdependence among component parts or members of a system seems to be taken for granted. Admittedly, individual systems operate according to their own internal rules of operation and thus on one level are closed to one another. But there is at the same time operative within systems theory the phenomenon of structural coupling, "a state in which two systems shape the environment of the other in such a way that both depend on the other for continuing their *autopoiesis* [self-constitution] and increasing their structural complexity."[39] Living systems (e.g., organisms—above all, those with a central nervous system and a brain) represent the necessary environment for psychic systems like the operation of the human mind; living systems and psychic systems in turn together provide the necessary environment for social systems (communities or various other forms of communication between and among human beings). So perhaps there is a way to incorporate systems theory within the scope of a new worldview or metaphysics based on principles of relationality rather than principles of substantiality, on principles of Becoming rather than principles of Being. After all, as Luhmann himself admits in his book *Social Systems*, there is need for a general systems theory that would legitimate a systems approach to biology, psychology, and sociology.[40] Such a general systems theory, to be sure, would be oriented to a commonality of method rather than a commonality of content: "General

---

37. Hans-Georg Moeller, *Luhmann Explained: From Souls to Systems* (Chicago: Open Court, 2006), 8.
38. Ibid., 3–21.
39. Ibid., 19.
40. Niklas Luhmann, *Social Systems*, trans. John Bednarz Jr. with Dirk Baecker (Stanford, CA: Stanford University Press, 1995), 12–16.

systems theory does not fix the essential features to be found in all systems. Instead, it is formulated in the language of problems and their solutions and at the same time makes clear that there can be different, functionally equivalent solutions for specific problems."[41] But is there in his notion of "self-referential systems," which critique their own operations as well as the operations of other systems,[42] a blend of contingency and necessity that seems to demand a metaphysical explanation? After all, as Etienne Gilson commented years ago in his book *The Unity of Philosophical Experience*, metaphysics has a way of burying its undertakers.[43]

In the chapter entitled "System and Function" in his book *Social Systems*, Luhmann begins by noting that while there are multiple types of real systems to be found in the world, his focus will be on self-referential systems, namely, "systems that have the ability to establish relations with themselves and to differentiate those relations from relations with their environment."[44] Instead of employing the conventional distinction between parts and wholes in his analysis of self-referential systems, Luhmann thus distinguishes between systems and their environments with the consequence that relations between and among entities within the system are more important than their relations with entities in the environment. As I will indicate below, such a definition of self-referential systems likewise seems to hold for Whiteheadian societies if they are considered as structured fields of activity for their constituent actual entities rather than simply as aggregates of individual actual entities with a similar pattern of self-constitution. Luhmann, to be sure, would be wary of this comparison because, for him, specifically social systems like those governing communities, organizations, and environments are "nonpsychic."[45] Their components are "elements" with objective relations to one another in virtue of the structure of the system;[46] they are not momentary subjects of experience with "internal" relations to one another.[47] Yet Luhmann also describes social systems as able to distinguish between themselves and their environment:

---

41. Ibid., 15.
42. Ibid., 487–88.
43. Etienne Gilson, *The Unity of Philosophical Experience* (Westminster, MD: Four Courts Press, 1982), 306.
44. Luhmann, *Social Systems*, 13.
45. Ibid., 14.
46. Ibid., 20–23.
47. Whitehead, *Process and Reality*, 58–59.

The theory of self-referential systems maintains that systems can differentiate only by self-reference, which is to say, only insofar as systems refer to themselves (be this to elements of the same system, to operations of the same system, or to the unity of the same system) in constituting their elements and their elemental operations. To make this possible, systems must create and employ a description of themselves; they must be able to use the difference between system and environment within themselves, for orientation and as a principle for creating information.[48]

Can a self-referential system make such decisions without some form of subjectivity for the system as a whole or some kind of intersubjectivity operative between the elements in their objective relations to one another?

Luhmann clearly wants to remain objective in his analysis of the workings of systems. Thus, while there are for him psychic systems, they coexist along with specifically social systems, organisms, and even machines within the ambit of systems theory.[49] But this means that human consciousness as a psychic system is a self-referential system, and as such, it is an "observer" of other self-referential systems insofar as these other self-referential systems are necessary parts of its environment.[50] Hence, while in Luhmann's view the concept of "subject" as used by Immanuel Kant and others should be replaced by the concept of self-referential systems,[51] the language of subjectivity is still present in his analysis of the workings of self-referential systems: "A system's internal organization for making selective relations with the help of differentiated boundary mechanisms leads to systems' being indeterminable for one another and to the emergence of new systems (communication systems) to regulate this indeterminability."[52] How does a system as a purely objective reality make "selective relations with the help of differentiated boundary mechanisms" without any form of internal self-awareness or subjectivity? Luhmann claims, "Selection can no longer be conceived as carried out by a subject, as analogous with action. It is a subjectless event, an operation that is triggered by establishing a difference."[53] But then he adds, "Difference does not determine what must be selected, only that a selection must be made. Above all, the system/environment difference seems to be what obliges the system to force itself,

---

48. Luhmann, *Social Systems*, 93.
49. Ibid., 2.
50. Ibid., 9, 16–17.
51. Ibid., 28.
52. Ibid., 29.
53. Ibid., 32.

through its own complexity, to make selections."[54] Here too, the language of subjectivity is once more present: the objective system/environment difference "obliges the system to force itself to make selections."

In his book *Luhmann Explained*, Hans Georg Moeller makes clear that Luhmann does not deny the de facto reality of human beings but only affirms that human beings exist on several levels at once (e.g., bodily, mentally, socially) and that these levels as autonomous self-referential systems do not make up an organic whole, a complete human being in the traditional sense.[55] Generalizing even further, Moeller argues that for Luhmann, "reality is not an all-embracing whole of many parts, it is rather a variety of self-producing systemic realities, each of which forms the environment of all the others. There is no common 'world' in reality, because reality is in each instance an effect of 'individual' systemic autopoiesis."[56] Luhmann consciously borrowed the term *autopoiesis* from Humberto Maturana and Francisco Varela, two biologists from Chile who applied systems theory to the study of biological reproduction, the way in which living cells are from moment to moment the product of their own internal processes of reproduction.[57]

Granted the usefulness of general systems theory as a common methodology for objective analysis in various otherwise loosely related scientific disciplines, one may still question whether one is thereby presented with an adequate understanding of human nature and the world of nature. Moreover, as Moeller comments in *Luhmann Explained*, the latter's understanding of systems theory as a "supertheory"[58] "does little outside of theory. With supertheory, the world does not become morally better, more rational, or spiritually complete. It only becomes more distinct."[59] So general systems theory, with its passion for objectivity, is an excellent tool for growth in knowledge but clearly inadequate for assisting human beings both as individuals and as members of society to live a better human life in greater harmony with the natural world. These latter goals would presumably be better attained by a worldview or metaphysics with a starting point in subjectivity—or even better, intersubjectivity—as the basis for moral activity as well as philosophical reflection. Yet such a worldview or metaphysics should also aspire to the same levels of objectivity as

---

54. Ibid.
55. Moeller, *Luhmann Explained*, 10.
56. Ibid., 14.
57. Ibid., 12–13.
58. Luhmann, *Social Systems*, 4–5.
59. Moeller, *Luhmann Explained*, 201.

Luhmann's systems theory. Hence, in the concluding pages of this chapter, I will briefly indicate how a Whiteheadian society when understood as a structured field of activity for its constituent actual entities generally corresponds to the need for objectivity in terms of systems theory and yet has its necessary grounding in the ongoing intersubjective relations of its constituent actual entities.

The Ontological Interdependence of Subjectivity and Objectivity

To begin, I repeat Luhmann's definition of self-referential systems as "systems that have the ability to establish relations with themselves and to differentiate these relations from relations to their environments."[60] In my view, this definition of a self-referential system also seems to fit the notion of a Whiteheadian society when understood as a structured field of activity for its constituent actual entities from moment to moment. Whitehead himself, of course, did not describe a society as a structured field of activity. But in his book *Process and Reality*, he says:

> Every society must be considered with its background of a wider environment of actual entities, which also contribute their objectifications to which the members of the society must conform. . . . But this means that the environment, together with the society in question, must form a larger society in respect to some more general characteristics than those defining the society from which we started. Thus we arrive at the principle that every society requires a social background, of which it is itself a part.[61]

If one presupposes that the terms "environment" and "field of activity" in this context are basically synonymous, then the environment / field of activity is in each case structured by the patterns of self-organization of its constituent actual entities in their ongoing succession. "Thus in a society, the members can only exist by reason of the laws which dominate the society, and the laws only come into being by reason of the analogous characters of the members of the society."[62]

Where I differ from Whitehead on this point is that he seems to derive the structural pattern for the environment / field of activity exclusively from constituent actual entities in terms of their status as "superjects," actual entities with a fully objectified pattern of self-constitution.[63] For many

---

60. Luhmann, *Social Systems*, 13.
61. Whitehead, *Process and Reality*, 90.
62. Ibid., 91.
63. Ibid., 27–28.

years now, I have argued that the patterns proper to the self-constitution of individual actual entities are necessarily incorporated into the governing pattern for the field of activity as a whole (the common element of form).[64] In this sense, my understanding of a Whiteheadian society corresponds closely to Luhmann's understanding of a system and its constitutive elements: "Elements are elements only for the system that employs them as units and they are such only through this system. This is formulated in the concept of autopoiesis."[65] That is, just as in Luhmann's understanding of systems and their elements, in my interpretation of Whiteheadian societies there is clear top-down causality from the common element of form of the society upon its constituent actual entities in their individual self-constitution from moment to moment. But whereas Luhmann, given his focus on objectivity, basically ignores the supporting role of individual elements in the formation of a system's governing structure, I agree here with Whitehead in his insistence that the ontological origin of the governing structure of the society comes from the ongoing interrelated activity of its constituents—actual entities as momentary self-constituting subjects of experience. Thus, whereas Whitehead in his understanding of a society focuses exclusively on the efficient causality of constituent actual entities in shaping their common element of form as a society, and while Luhmann emphasizes the formal causality of the governing structure of the system in organizing its various elements, I choose the middle path in my claim that a Whiteheadian society and a self-referential system for Luhmann should be considered as constituted in equal measure by bottom-up efficient causality and by top-down formal causality. In this way, there is a suitable combination of subjectivity and objectivity in producing the functional unity of either a Whiteheadian society or a self-referential system for Luhmann.

Still another feature of a self-referential system as described by Luhmann in *Social Systems* is to found in his notion of system differentiation: "System differentiation is nothing more than the repetition of system formation within systems. Further system/environment differences can be differentiated within systems. The entire system then acquires the function of an 'internal environment' for these subsystems, indeed for each subsystem in its own specific way."[66] This can be usefully compared with Whitehead's notion of "structured society," a society "which includes subordinate societies and nexuses with a definite pattern of structural interrelations. . . . A structured society as a whole provides a favorable environment for the

---

64. Ibid., 34.
65. Luhmann, *Social Systems*, 22.
66. Ibid., 18.

subordinate societies which it harbours within itself. Also the whole society must be set in a wider environment permissive of its continuance."[67] Luhmann's notion of system differentiation and Whitehead's understanding of structured societies, however, are brought into even closer conceptual alignment if one thinks of both Whiteheadian societies and Luhmann's self-referential systems in terms of structured fields of activity for their constituent elements or constituent actual entities. Physical reality, in other words, is best seen in terms of fields within fields. Yet each field or system possesses an internal unity and thus has an individual identity by reason of the structural principles proper to itself, even as it contributes to the structure of fields or systems more comprehensive than itself.

What is to be said, however, about an ultimate or all-inclusive field of activity? For Whitehead, this ultimate, all-inclusive field of activity is the consequent nature of God, God's ongoing experience of the world as a whole in which "the revolts of destructive evil are dismissed into their triviality of merely individual facts, and yet the good they did achieve in individual joy, in individual sorrow, in the introduction of needed contrast, is yet saved by its relation to the completed whole."[68] For me, as one who believes in the Christian doctrine of the Trinity, the ultimate and all-inclusive field of activity is the kingdom of God, the participation of all creaturely actual entities and the societies to which they belong in the divine field of activity, the communitarian life of the three divine persons. But as Moeller points out in *Luhmann Explained*, for Luhmann the global society is not synonymous with a harmonious whole:

> Global society is a complex multiplicity of subsystems, which are not integrated into an overarching global unity. Function systems [e.g., the natural and social sciences, economics, international politics] operate beyond geographical borders; in this sense they are universal. There is no geographical space where they cannot go, but at the same time they are all functionally particular. They are bound by their function, not by space. Global society consists of a plurality of systems that are both universal and particular.[69]

So in the end Luhmann, as a secular thinker with strong affinities for postmodernism and French deconstructionism, stands apart from Alfred North Whitehead and me. Both of us are metaphysicians with strong beliefs in the classical notion of *cosmos*—or better, *chaosmos*—an organic but

---

67. Whitehead, *Process and Reality*, 99.
68. Ibid., 346.
69. Moeller, *Luhmann Explained*, 54.

still developing totality or open-ended system.[70] Both of us, to be sure, share with Luhmann an evolutionary approach to reality, but we disagree that a functional, systems-oriented approach to reality can dispense with human subjectivity as a necessary starting point for an explanation of how evolution works everywhere in the world of nature.

But while Whitehead is preoccupied with subjectivity in terms of actual entities (momentary self-constituting subjects of experience) and the societies into which they normally aggregate, my focus in this part of chapter seven has been on the construction of a mediating position between subjectivity (individual actual entities) and objectivity (the societies that transmit the persistent pattern of interrelation of these actual entities from moment to moment). As noted earlier, subjectivity and objectivity are intrinsically interrelated. Subjectivity only operates within the context of an already existing state of affairs. Yet change in this objective state of affairs only happens through the ongoing interplay of its subjective components.

---

70. See Michel Weber, *Whitehead's Pancreativism: Jamesian Applications* (Frankfurt am Main: Ontos Verlag, 2011), 209: "We live in a partially ordered world where stability is earned over creative processes, where stability never has the last word."

8

The Democratic Process as an Open-Ended System in Political Life

In Western Europe and North America, democracy has become in recent centuries the most common form of political government because in principle it guarantees the rights of individual citizens against various forms of tyranny and oppression. Likewise, when in various parts of the world a totalitarian regime is overthrown, the notion of a democratic system of government is regularly set forth by reformers as the ideal toward which to aspire in setting up a new regime. One need only think of the overthrow of totalitarian regimes in the former Soviet Union in the twentieth century and the current unrest in traditionally one-man or one-party governments in the Middle East and in the Far East. But is democracy in the end wishful thinking, something more often discussed in theory than achieved in practice? Or does it have legitimation in the very structure of physical reality such that one can appeal to it as the most "natural" form of human government, even if in practice it is not always effectively implemented? It will be the purpose of this chapter to argue that the second alternative is, or at least should be, the case. I will also contend, however, that until recent centuries this was not the case, since the bias of educated people in Western society and elsewhere in the world was toward monarchical or at least aristocratic forms of government based, in part, upon implicit antecedent philosophical convictions about the underlying nature of reality. Hence, what is currently needed to reinforce the practical conviction that democracy is indeed the most suitable form of government for human beings is a metaphysical conceptuality that will effectively challenge the older thought pattern that inclined one unconsciously toward monarchical or aristocratic forms of government.

Fortunately, work in this direction has already been done by Colin Gunton in his book *The One, the Three and the Many*.[1] Therein he argues that the "pathos of the modern condition" is due in large measure to the uncritical acceptance of an outdated understanding of the relationship between the One and the Many, ultimately derivative from the philosophy of Plato and Aristotle, whereby the One (or, in any case, the Few) are entitled to hegemony over the Many. Even modern efforts to displace the One in favor of the Many have resulted in new forms of totalitarianism since a new paradigm for the relationship of the One and the Many that should favor more democratic forms of government is not yet in place within contemporary Western culture.[2] His solution is to have recourse to the Christian doctrine of the Trinity among the early Greek fathers of the church as a mutual indwelling or *perichoresis* of the three divine persons so as to constitute the reality of one God. While I share with Gunton the belief that the Christian doctrine of the Trinity when understood as a community of divine persons has enormous value for this new understanding of the relationship of the One and the Many, I also believe that one need not believe in the doctrine of the Trinity to deal with the strictly philosophical issue of a change of paradigm for one's basic understanding of the relationship between the One and the Many and its importance for legitimating the move toward more democratic forms of government in modern life. Accordingly, I will first review Gunton's critique of the classical understanding of the relationship between the One and the Many with attention to the negative effects that this habitual thought pattern has had on Western governments. Afterward, I will make clear how my own rethinking of the notion of "society" in the philosophy of Alfred North Whitehead could provide theoretical justification for the new, more democratically conceived understanding of the One and the Many, which Gunton claims is indispensable for remedying the "pathos of the modern condition." Not a reconceived doctrine of the Trinity, therefore, but a philosophical claim based upon a rethinking of Whitehead's notion of society will be the centerpiece of my argument that democracy is the most "natural" form of government for human beings since, at least in principle, it corresponds more closely to the nature of reality.

Gunton begins his analysis of the ills of Western civilization with the pre-Socratic philosophers, notably Heraclitus and Parmenides, whose views on Ultimate Reality are polar opposites:

---

1. Colin E. Gunton, *The One, the Three and the Many: God, Creation and the Culture of Modernity* (Cambridge: Cambridge University Press, 1993).
2. Ibid., 34–40.

> Heraclitus is the philosopher of plurality and motion: the many are prior to the one, and in such a way that there is to be found in nature no stability. Parmenides represents the opposite pole of thought. . . . Reality is timelessly and uniformly what it is, so that Parmenides is the philosopher of the One *par excellence*. The many do not really exist, except it be as functions of the One.[3]

According to Gunton, Plato, the most influential of the ancient philosophers, privileged the philosophy of Parmenides over that of Heraclitus. That is, he distinguished between the unchanging world of the intelligible Forms and the changing world of the senses in which material things provisionally embodied the Forms in an imperfect manner.[4] Implicitly, then, Plato's social and political thought exhibits "a strong tendency to totalitarianism, not in the modern sense, but in the sense of a preference of the one to the many, of unity to diversity."[5]

As we have already seen in chapter five, early Christian philosophers and theologians like Thomas Aquinas transposed this paradigm for the relationship of the One and the Many into their understanding of the God-world relationship. That is, the Platonic Forms were thought to exist eternally in the mind of God who, as the Creator of the material universe, is thus the transcendent One ordering to Himself the empirical Many of the physical universe. In this way, a basically monarchical understanding of the relationship between the One and the Many was uncritically incorporated into the Christian theology of the Middle Ages. Since God is clearly transcendent of the world that God creates, then the One must be transcendent of the Many as their necessary principle of unity and order. Likewise, monarchy was the preferred form of government in the Middle Ages because it was thought to reflect the nature of reality, the subordination of material reality to its transcendent Creator and all-wise Ruler.

Beginning with Descartes in the early modern period, this paradigm for the relationship between the One and the Many was partly challenged and partly reinforced. As Gunton comments, it was challenged insofar as God was no longer conceived as the transcendent One. Instead, the "unifying rational mind" of the individual human being became by degrees "the seat of rationality and meaning" for one's understanding of the physical

---

3. Ibid., 17–18.
4. Plato, *The Republic*, bk. VI, nos. 510–11, in *The Collected Dialogues of Plato*, ed. Edith Hamilton and Huntington Cairns (Princeton, NJ: Princeton University Press, 1989), 745–47.
5. Gunton, *The One, the Three and the Many*, 21.

world.[6] At the same time, the primacy of the One over the Many was reinforced in that the individual was implicitly encouraged to think of all the other persons and things of this world as instrumental to the satisfaction of his or her interests and desires. Such a situation in the social and political sphere, of course, cannot long perdure, since it leads to a state of continual conflict such as Thomas Hobbes describes in *Leviathan*.[7] In the end, for the sake of peace and social order, individuals surrender their individual freedom to a monarch who imposes an arbitrary rule of law to which all must conform.[8] Thus, early modern efforts to give priority to the Heraclitean Many over the Parmenidean One ended, in Gunton's judgment, in failure because they simply replaced God as the transcendent One with an immanent One far more tyrannical in its behavior toward the empirical Many than the God of medieval theology.[9]

This is not to deny, of course, that since the time of Thomas Hobbes, progress has been made in setting forth principles of democratic government. John Locke with his *Second Treatise on Civil Government* and Jean Jacques Rousseau with his book *The Social Contract* have set forth guidelines for the establishment of working democracies, such as the United States of America, around the world. But as even a superficial analysis of the writings of these two distinguished philosophers makes clear, there is still ambiguity in the relationship of the citizens to one another and to the state or the civil community. Locke, for example, stipulated in the above-mentioned treatise that a community acts as one body "through the will and determination of the majority."[10] But as James Collins comments, while Locke's proviso counteracts the "idolatry" of the state such as found in Hobbes's theory, "it does not meet the need for well-determined safeguards against a tyrannous majority, with whom the political power ultimately lies."[11] Rousseau, on the other hand, in his *Social Contract* distinguished between the will of all (*volonte de tous*) and the general will

---

6. Ibid., p. 28.

7. Thomas Hobbes, *Leviathan, or the Matter, Forme and Power of a Commonwealth, Ecclesiastical and Civil*, ed. Michael Oakeshott (Oxford: Blackwell, 1960), I, 13.

8. Ibid., II, 17.

9. Gunton, *The One, the Three and the Many*, 33.

10. John Locke, *An Essay Concerning the True Original, Extent and End of Civil Government*, chap. 8, n. 96, in *The English Philosophers from Bacon to Mill*, ed. Edwin Burtt (New York: Modern Library, 1939).

11. James Collins, *A History of Modern European Philosophy* (Milwaukee, WI: Bruce Publishing Co., 1954), 362.

(*volonte generale*).[12] The former is, according to Rousseau, simply the sum of the particular wills in a community as manifested in a majority, or even a unanimous, vote. The latter, on the contrary, is to be understood as the genuine will of the community as a whole, which represents the true will of the individual citizen, even though it may militate occasionally against his or her instinctive desire or self-interest.

Rousseau can be criticized for the obscurity of his notion of the general will on at least two counts. First, he provides no criterion whereby a given decision of a legislative body can be evaluated as the general will of the community or simply as the particular will of a self-seeking majority of its citizens. Second, an individual citizen, according to Rousseau, can be constrained to obey the general will of the community on the presupposition that the general will expresses his or her own deeper desire. He or she is, so to speak, "forced to be free."[13] On both grounds, Rousseau can be accused of conceding to the state totalitarian power over its subjects. At the same time, Rousseau, far better than Locke, recognized that a community is more than simply the aggregate of its members, that it is a specifically social reality with goals and values proper to itself that are somehow embodied in the "general will" of the group at this particular moment in history. Difficult as it may be, therefore, to decide whether a given proposal truly corresponds to the general will of the members, a community cannot long survive without a general consensus among those same members as to their basic group identity and group purpose—in other words, their general will.

As indicated in previous chapters of this book, I use a revised understanding of the category of society in the philosophy of Alfred North Whitehead to indicate how a specifically social reality can arise out of the dynamic interplay of its individual parts or members. In this chapter I contend that this revised understanding of a Whiteheadian society is very useful for justifying philosophically the validity and continued use of the democratic process as a decision-making tool in modern political life. To make that clear, I first review briefly Whitehead's understanding of the nature of society and my revision of that concept. According to Whitehead, the members of a society, actual entities, "prehend" the pattern of interrelation existent among their predecessors in that same society and by a process of "transmutation" incorporate it into their own individual processes of concrescence so as to perpetuate the "common element of form" for

---

12. Jean Jacques Rousseau, *The Social Contract*, bk. 2, chap. 3, in *The Social Contract and Discourses*, trans. G. D. H. Cole (London: Everyman's Library, 1923).
13. Ibid., bk. 1, chap. 7.

their existence as that society here and now.[14] Thus, while individual sets of actual occasions come and go, the society remains as a self-sustaining reality with a distinctive social identity or pattern of interaction among its constituent members from moment to moment. The problem is that a Whiteheadian society thus appears to be nothing more than an aggregate of individual entities genetically related to one another in terms of an analogous pattern of individual self-organization. Moreover, since there is necessarily a new aggregate at every moment with each new set of actual occasions, then, contrary to what Whitehead himself says in *Process and Reality*, a society is not self-sustaining.[15] It too comes and goes with each new moment of the cosmic process. Whitehead, to be sure, seems to anticipate this objection with his further comment that in his philosophy "it is not 'substance' which is permanent, but 'form.'"[16] Form carried over from one aggregate of actual occasions to another is the principle of continuity in a changing world rather than substance, an immaterial substratum for ongoing accidental changes in the physical order. But where is the form located as a uniform, objective reality that can be transmitted from one set of actual occasions to another? If it is only analogously present in each constituent actual occasion of a given society here and now and if the next set of actual occasions must "transmute" that form so as to constitute it as the basic pattern of interrelation for their own coexistence as the same society in the following moment, then the objective continuity of form for the ongoing existence of the society as a self-sustaining reality would seem to be constantly in jeopardy. What is to guarantee that the society will be basically the same objective reality from one moment to the next?

For these reasons, I have argued at length in this book that a Whiteheadian society is an objectively constituted environment or ongoing structured field of activity for successive sets of actual occasions. There is certainly support for my contention in Whitehead's own remarks in *Process and Reality*: "Every society must be considered with its background of a wider environment of actual entities, which also contribute their objectifications to which the members of the society must conform. . . . But this means that the environment, together with the society in question, must form a larger society in respect to some more general characters than

---

14. Alfred North Whitehead, *Process and Reality: An Essay in Cosmology*, corrected edition, ed. David Ray Griffin and Donald W. Sherburne (New York: Free Press, 1978), 34, 250–54.

15. Ibid., 89: "The point of a 'society,' as the term is here used, is that it is self-sustaining; in other words, that it is its own reason."

16. Ibid., 29.

those defining the society from which we started."[17] For Whitehead's term "environment" I substitute "structured field of activity." As I see it, this substitution allows me to claim that a Whiteheadian society, precisely as an environment or structured field of activity, has an objective reality over and above its constituent actual entities from moment to moment. The field, in other words, is the bearer of the common element of form linking successive sets of actual entities so as to constitute them as a self-sustaining reality over time. It is, to be sure, derivative from the genetic relation of successive actual entities or sets of actual entities, but it survives as these actual entities arise and perish.[18]

Yet even given the logical coherence of my position here on the nature of Whiteheadian societies, there remains a subjective or psychological hurdle to its acceptance by many, if not most, people. It is easy to think of the world as made up of "things," both living and nonliving. But to think of the world as made up of structured fields of activity that overlap one another horizontally and are hierarchically (vertically) ordered, one inside another, is a major psychological obstacle. It presents perhaps an even bigger problem for understanding and acceptance than Whitehead's own proposal that the world is ultimately made up of "spiritual atoms" rather than material atoms, momentary self-constituting subjects of experience rather than tiny bits of matter being pushed and pulled in different directions by the laws of gravity and electromagnetism. But as I shall explain further in chapters nine and ten, commonsense experience is not a good guide for understanding what is going on both within us and around us. What we see, hear, smell, taste, and touch does not precisely correspond to physical reality apart from our perception of it. For if our sense perception took in everything going on within us and around us, we would be overwhelmed by the enormous wealth of empirical data, which would need to be classified and reduced to some manageable order. So, as indicated already in chapter six, the subconscious workings of our minds and bodies have been antecedently shaped by natural selection so that our ancestors would have a better chance to survive and even prosper in an impersonal, at times even hostile, natural world.

Yet among natural scientists, the notion of field as at least a methodological presupposition of the organization of empirical data is fairly common, and for the more philosophically oriented members of the scientific community, the notion of field has been gaining acceptance as likewise an

---

17. Ibid., 90.
18. Cf., for example, my book *The One in the Many: A Contemporary Reconstruction of the God-World Relationship* (Grand Rapids, MI: Eerdmans, 2001), 131–55.

ontological principle, something descriptive of physical reality apart from the workings of the human mind. As Leemon McHenry comments in a recent article, Albert Einstein "viewed Faraday and Maxwell's theory [of electromagnetism] as the greatest change in the axiomatic basis of physics and in our conception of the structure of reality."[19] Likewise, in the same article McHenry cites Willard Quine: "Matter is quitting the field, and field theory is the order of the day."[20] To make clear how this notion of field is also gaining ground among philosophers of science, I turn now to a recent book by Ervin Laszlo, one of the early systems-oriented thinkers to whom I made reference in chapter five. Laszlo, in *The Connectivity Hypothesis*, published in 2003, seems to be thinking of natural systems, using field imagery.[21]

Drawing upon contemporary research in the field of quantum physics, Laszlo postulates the existence of a primordial energy field coordinating the interaction of energy-events or "particles" at the quantum level so that they remain "entangled" even when separated in space and time beyond the limits of conventional cause-effect relationships. Furthermore, at the molecular, organic, and supraorganic levels of nature, one finds "nonconventional" connections between the parts that make up a system and between the systems and their environment. He initially concludes, "space is not a vacuum but a plenum, and information, as physically effective 'in-formation,' is as fundamental as energy, and is likewise conserved."[22] To support that claim, he marshals evidence from quantum physics, cosmology, the life sciences, and transpersonal psychology. In each case, he notes that the alleged system-wide correlations between parts and wholes seem to be operative even among parts widely separated from one another in space and time. He then sets forth four operative principles underlying his new worldview:

> Anomalous coherence in a system implies quasi-instant correlation among the parts and components of that system.

---

19. Leemon B. McHenry, "Maxwell's Field and Whitehead's Events: The Adventure of a Revolutionary Idea," in *Subjectivity, Process, and Rationality*, ed. Michel Weber and Pierfrancisco Basile (Frankfurt am Main: Ontos Verlag, 2007), 1978.

20. Ibid. See also Willard Quine, "Whither Physical Objects?" in *Essays in Memory of Imre Lakatos*, ed. R.S. Cohen, P. K. Feyerabend, and M. W. Wartofsky, Boston Studies in the Philosophy of Science 39 (Dordrecht: D. Reidel, 1976), 497–504.

21. Ervin Laszlo, *The Connectivity Hypothesis: Foundations of an Integral Science of Quantum, Cosmos, Life, and Consciousness* (Albany, NY: State University of New York Press, 2003).

22. Ibid., 1–2.

Such correlation implies system-wide connectivity.

System-wide connectivity implies in turn the presence of an interconnecting medium.

In a realist perspective the interconnecting medium is a system-wide field.[23]

This system-wide field, as already noted, is an energy-filled plenum, a cosmic plenum (as opposed to simply a cosmic vacuum) whose short-term "virtual" energies are in ongoing interaction with the particles and systems of particles making up the visible universe. As a result, Laszlo sees physical reality as unitary but, for the purpose of analysis, divided into two principal domains:

> One is the manifest domain of (directly or instrumentally) observable particles and systems of particles; the other the virtual domain of the cosmic plenum, the energy sea from which the particles arise, with which they interact, and into which they ultimately fall back. The latter domain is intrinsically unobservable, but it is inferable through its effects on the observable domain. . . . The interaction of the two domains generates the observable entities—the particles and systems of particles—of the universe.[24]

Particles and systems of particles within the manifest domain are not in the strict sense material realities but in reality "vibrating nodal points (distillations or crystallizations) of the energies of the virtual domain."[25] As such, they bear in my judgment a distinct resemblance to Whitehead's notion of actual entities, momentary self-constituting subjects of experience, and their simultaneous aggregation into societies of various degrees of complexity.

Besides the notion of "field," the other key concept in Laszlo's scheme is the notion of "in-formation" which he explains as follows:

> The two-way interaction between systems and plenum is not a zero-sum exchange of the same information cycling back and forth, for the wave function encoded in the psi-field [cosmic plenum] is that of the *collective* state of the entities that created it. . . . Thus when the plenum response carries the wave function of a set of particles, it in-forms those particles with the collective state of the higher-level coordinate system in which they participate.[26]

23. Ibid., 40.
24. Ibid., 104.
25. Ibid.
26. Ibid., 75.

This seems to correspond well with my own contention that a Whiteheadian society is a field of activity for its constituent actual occasions that is structured by the dynamic interplay of those same actual occasions. The common element of form for the society, accordingly, resides primarily in the structure of the field rather than in the self-constitution of the individual, constituent occasions, taken one by one. As Laszlo comments, the wave function proper to the system "in-forms" the particles making up the system. The particles remain distinct from one another, but together they constitute a social totality greater than (and to some extent other than) themselves as individuals.

As I also noted in chapter five, Laszlo believes that nature is in this way hierarchically ordered; that is, it is made up of "a nested hierarchy of nonlocally connected coherent systems."[27] Lower-level systems exist as semiautonomous realities within higher-level systems, thereby setting limits to the function of the higher-level systems, just as the higher-level systems provide additional "in-formation" to the operation of the lower-level systems. Thus, the universe as a whole and its innumerable subfields of activity are nonlocally connected with one another, but in such a way as to allow relative autonomy of existence and activity both to the various subsystems and to their constituent parts or members. Whitehead seems to have the same hierarchical vision of reality when he notes that "there is no society in isolation. Every society must be considered with its background of a wider environment of actual entities, which also contribute their objectifications to which the members of the society must conform."[28] Thus, for Whitehead as well as for Laszlo, there is clearly "upward" causation in terms of the social background of which each society is a part. Downward causation, or what Laszlo calls the "in-formation" provided to lower-level systems by higher-level systems, is more implicit than explicit in Whitehead, largely because, as noted above, he did not adequately think through the relationship of societies to one another as well as of actual entities to one another. But something akin to downward causation seems to be at work in his analysis of "structured societies" in *Process and Reality*. He comments, "A structured society as a whole provides a favorable environment for the subordinate societies which it harbors within itself."[29] What can this mean except that the structured society is governed by a common element of form that has considerable impact upon its subsocieties in and through

---

27. Ibid., 1.
28. Alfred North Whitehead, *Process and Reality*, 90.
29. Ibid., 99.

the constituent actual entities of those same subsocieties that prehend that common element of form pervading the structured society as a whole?

In any event, the field metaphor common to Laszlo's scheme and my own reconstruction of Whitehead's metaphysics allows one to picture both life on this earth and the entire visible universe in terms of hierarchically ordered fields of activity, each with its component parts or members governed by specific laws or patterns of behavior proper to the field as such. But what does all this speculation about the field-oriented structure of physical reality have to do with the question posed at the beginning of this chapter, namely, whether democracy is the most natural form of political organization among human beings? If we bear in mind that a Whiteheadian society or a system in Laszlo is a structured field of activity for its constituent parts or members, and that the prevailing structure of that field of activity first comes into being and then is sustained in existence only in virtue of the coordinated activity of its constituent parts or members (ultimately, actual entities as self-constituting subjects of experience), then in effect every form of political organization for human beings is, at least in principle, democratic. That is, even if the human beings in question choose to be governed by a king or queen, the monarch exercises power over his or her fellow human beings only in virtue of their explicit or at least implicit assent. That is, there is no "divine right of kings" allowing for a succession of monarchs by divine will and authority. Moreover, even though the agency proper to the state under these circumstances is exercised principally but not exclusively by the monarch, the agencies proper to the individual citizens are also required for the proper functioning of the political community as a genuinely corporate reality.

Thus in terms of the new paradigm for the relationship between the One and the Many set forth in this book, one must say that the unity of the state or civil community is emergent out of the interplay of the many members of the state with one another according to various patterns of existence and activity. It is not simply the monarch or the ruling body within the state that, as the transcendent One within the totality, gives order and direction to the otherwise chaotic activities of the Many. Rather, as noted above, the One (in this case, the state, whether constituted as a monarchy, aristocracy, or democracy) is at every moment the byproduct or result of the activity of the Many (the citizenry) vis-à-vis one another. Using the field-oriented understanding of societies or systems proposed in this chapter, the state or civil government is a wide-ranging structured field of activity for its citizenry, which endures as individual citizens are born, live their lives, and eventually die. But the state is itself dependent for its own existence and ever-changing political, economic, and cultural structure upon those same citizens and their sustained interplay with one another over time. The state, of course, is not the only structured field of activity proper to its citizens in their private lives. Individuals are inevitably gathered into many other forms of community (economic, religious, purely social) that also are structured fields of activity for their members. But in terms of the political life of its citizens, the state (local, regional, national, and international) is invariably a structured field of activity dependent upon its citizens for its own continued existence in one form or another.

Thus, while the classical paradigm for the relationship between the One and the Many implicitly legitimated various forms of totalitarianism on the grounds that individual citizens were incapable of peaceful coexistence apart from the conditions of law and order imposed by the state, this new paradigm for the relationship between the One and the Many is radically democratic in that it invests the ultimate authority for the regulatory power of the state in the consent of its citizenry. This is not to deny, of course, that a strong executive branch of government is needed to implement the will of the people in a representative democracy once that becomes known through the legislative process. But for that same reason, a civil constitution and an independent judiciary are also needed to curb the power of the executive and the legislative branches of government when either or both of these juridical entities are exceeding their constitutionally guaranteed jurisdiction. In short, the state or civil community is not a transcendent Platonic ideal with predetermined structures and procedures but an evolutionary reality emergent out of the ongoing interplay of the various popular constituencies within it. Admittedly, the state with its current legal structures and its various public institutions is likewise the basic principle of continuity in space and time for those same interactions among its citizens; yet it is still capable of gradual evolution both in its legal structure and in its range of activity in response to the manifest will of the people as time goes on.

Here, of course, one can legitimately ask what is meant by "the manifest will of the people." Is it simply the will of the numerical majority here and now on a given issue? If so, then the logical problems associated with the political philosophy of John Locke and Jean Jacques Rousseau, as mentioned above, recur in still another form. My response would be that a Whiteheadian society best functions when the individual agencies proper to all its constituent actual occasions are suitably coordinated and directed to the achievement of a common goal or purpose. It functions less well to the extent that these individual agencies are somewhat in conflict with one another, and it ceases to exist as a society if and when these agencies are seriously in conflict with one another. Hence, at least in principle, a Whiteheadian society cannot long endure when its constituent actual occasions are in mutual disharmony. From that perspective, the notion of a Whiteheadian society allows one to move beyond Locke's purely quantitative assessment of the will of the group in terms of a numerical majority and to approximate what Rousseau had in mind with the notion of the "general will."

Yet even better than Rousseau, one has at hand an informal mechanism for measuring that general will and for enhancing it once it starts to take hold within a given population. That is, as Whitehead points out with respect to societies in general, actual occasions constitutive of a society are linked together by their joint prehension of a common element of form or ideal pattern of behavior that has to be internalized in its own way by each of the actual occasions. The greater the affinity between the actual occasions in this matter, the stronger the unity of the society in question will be. Hence, a civil community will be stronger and more stable, the more its individual citizens consciously share a common heritage and look forward to a common destiny.[30] Efforts by one interest group within the state to systematically repress the legitimate interests and desires of other groups, on the contrary, will inevitably lead to civil unrest and possibly open violence as the repressed groups seek a justifiable redress of grievances. Once again, however, it is important to realize that democracy as the preferred form of civil government always remains more an ideal than an already achieved reality. As the U.S. government has painfully experienced over the last few decades in dealing with various foreign governments around the world, setting up allegedly free elections for a new government, even with official foreign observers on hand, does not guarantee satisfactory results. A certain level of maturity on the part of the citizenry and, above all, their elected officials is needed to guarantee attention to the common good as well as to personal and group interest. For that matter, one can question whether the democratic process here in the United States, "the land of the free and the home of the brave," is still functioning as originally planned or has fallen prey to power interests with a virtually unlimited supply of money to guarantee legislation in line with their own goals and values.

Yet even with these reservations about the workings of a Whiteheadian society in the political order, it still seems safe to claim that, pace Luhmann,[31] subjectivity—or more precisely, intersubjectivity—plays a key role in first originating and then sustaining the system or network of

---

30. Cf. Josiah Royce, *The Problem of Christianity* (Chicago: University of Chicago Press, 1968), 248–49.

31. Cf. chapter seven.

subsystems constituting a representative democracy. To be sure, no democratically organized form of government will remain unchanged in its basic mode of operation for any significant length of time. For as a corporate reality emergent out of the ever-changing interactions of human beings with one another, democracy, like any other open-ended system, will over time inevitably change character and move now in one direction, now in another. Whether these changes will be for better or for worse will only be decided in retrospect as a consequence of how things have concretely worked out. This is, after all, both the challenge and the risk of rethinking the relation between the One and the Many in the political order, so that the Many (the citizenry) are the ultimate agents of change, and the One (the civil government in its constitutional structure and normal mode of operation) is the corresponding principle of continuity.

Part Three

Christian Doctrinal Questions

9

Incarnation and Redemption within the Cosmic Process

At the Council of Nicaea in 325 CE, the church fathers declared that Jesus Christ as the incarnate Son of God was "of one substance" with the Father and thus divine as well as human.[1] But only at the councils of Ephesus and Chalcedon in 431 and 451 CE, respectively, was it made clear that Jesus as the incarnate Son of God is a divine person with two natures, one divine and the other human. The divine nature does not absorb the human nature, nor does the human nature fully encompass the divine nature. Each nature is distinct and yet is inseparably united with the other.[2] Christ as a divine person thus has a true human soul, has a real human intellect and will, and performs genuinely human actions. Beyond this definition of terms, the church fathers wisely did not venture. Instead, they left to subsequent generations of theologians down to the present day the job of further explaining how one and the same person can, without confusion, exist within and effectively exercise two natures that are so utterly different from one another.

In this chapter I will offer a process-oriented explanation of how Jesus is both divine and human at the same time. My explanation will be based on the panentheistic model of the God-world relationship, in which the three divine persons of the Christian doctrine of the Trinity and all their creatures share a common space for their life together; together they structure that common space by their individual decisions from moment to

---

1. *Enchiridion Symbolorum: Definitionum et Declarationum de rebus fidei et morum*, ed. Henricus Denziger and Adolfus Schönmetzer, SJ (Freiburg: Herder, 1973), no. 125.

2. Ibid, 301–2.

moment. Within that frame of reference, the doctrine of the incarnation is no longer a huge speculative puzzle. For if every created subject of experience keeps its own individuality, its own finite field of activity, even as it collaborates with the three divine persons in giving form or structure to a common field of activity (the ongoing kingdom of God), then the claim that Jesus as the incarnate Son of God functions in two distinct fields of activity, one divine and the other human, which sufficiently overlap so as to create a common divine-human field of activity, is really not all that remarkable. Instead, the problem may be how to distinguish Jesus as the incarnate Son of God from all other created subjects of experience who have an intersubjective relationship to the three divine persons.

My answer to this objection is that, unlike other created subjects of experience, Jesus in his human consciousness not only experienced union with the three divine persons in a general way but experienced in a special way an ontological unity with just one of those persons, the divine Word, who is the self-expression of the Father within the divine life. Thus, the difference between a psychological *union* between two or more separate subjects of experience and the ontological *unity* of two interdependent subjectivities or psychic agencies within one person is what makes Jesus in his human nature completely different from all other created subjects of experience.[3] Whereas all other created subjects of experience feel, in varying degrees, the presence of God as other than themselves and thus can only aspire to greater union with God, Jesus felt the presence of the divine Word in his human consciousness to be the same as himself. That is, during his public ministry, when he spoke and acted, he felt himself to be in fact God's Word, the Father's emissary here and now, to the people of his generation. Furthermore, as the Gospel writers commented, his hearers invariably caught Jesus' extraordinary sense of self-assurance, his unique self-identity, since they marveled that he spoke with such authority, not like the scribes and Pharisees (e.g., Matt 7:29).

From a philosophical perspective, however, how is this to be explained? I believe that my reinterpretation of Whitehead's understanding of the God-world relationship can provide us with at least a plausible explanation. In his analysis of actual entities, momentary self-constituting subjects of experience, Whitehead claimed that a self-constituting actual entity is always guided in its "concrescence" by what he called its "subjective

---

3. Joseph A. Bracken, *The Triune Symbol: Persons, Process and Community* (Lanham, MD: University Press of America, 1985), 53.

aim," its feeling-level sense of purpose in coming to be itself.[4] Strongly influencing that subjective aim, of course, would be first the decisions of its predecessors within the "society" of actual entities to which it belongs and then the particular circumstances of its environment, the world around it here and now. But, said Whitehead, still another important factor in its subjective aim here and now is what I have referred to earlier as a divine "initial aim,"[5] equivalently a moment of divine grace in which God the Father both enlightens and motivates a given actual entity to make its self-constituting "decision" one way rather than another.

While this divine initial aim is active in the self-constitution of all actual entities, even those which are the constitutive parts of inanimate things such as tables and chairs, within the consciousness of human beings the divine initial aim is equivalently the voice of conscience. Naturally, the actual entity is free to accept or reject or partially modify this inspiration from God in line with other feeling-level factors in making its decision here and now. What we can presume on the basis of our Christian belief in the sinlessness of Jesus, however, is that Jesus, unlike other human beings, never seriously deviated from the Father's initial aim for him at any moment of his adult life.[6] In this way, he was effectively one person with the divine Word in the Word's own relationship to the Father within the divine community. That is, under the inspiration of the divine initial aim at every moment, Jesus in his human nature offered himself to the Father and the Spirit with the same spirit of self-giving love as the divine Word did in dealing with the other two divine persons in their communitarian life together. As a result, Jesus was unconsciously one person with the divine Word in dealing with the Father and the Spirit.

Before his resurrection, of course, Jesus' relationship to God must have been more on a feeling level than on a strictly cognitive level. Otherwise, it is hard to imagine him as being truly human as well as divine. Along the same lines, one may conjecture that Jesus only realized his ontological identity with the divine Word—and in that sense, his divinity—at the resurrection, when the inevitable limitations of conscious life in the body were somehow set aside. Here one could object that Jesus' mother, Mary, is likewise believed to have been sinless from the moment of her conception and yet has never been considered divine by Christians. No one claims

---

4. Alfred North Whitehead, *Process and Reality: An Essay in Cosmology*, corrected edition, ed. David Ray Griffin and Donald W. Sherburne (New York: Free Press, 1978), 25.

5. Ibid., 244.

6. Bracken, *The Triune Symbol*, 70.

that she is one person with either the Father or the Spirit, still less with the person of the Word. But this is to overlook that in Whitehead's explanation of divine initial aims, they are not seen as purely generic, aimed at actual entities in general, but rather as quite specific, tailored to the particular situation of individual actual entities within the society to which they belong at any given moment. Hence, while God the Father's initial aims to Mary, at least from the annunciation onwards, were presumably in line with her role as the mother of the Messiah, the Father's initial aims to Jesus as the Messiah or the Word Incarnate would be completely different, enabling Jesus but not Mary to be one person with the divine Word within the life of the Trinity.[7]

But theory aside, how in actual practice would this work out both for the divine Word and for Jesus in his human consciousness? Quite honestly, no one really knows. But we can conjecture that by becoming one person with Jesus of Nazareth, the divine Word experienced all the joys and sorrows, anxieties and uncertainties of Jesus in his human consciousness and identified with them as his own. At the same time, in virtue of his divinity, the divine Word was able to put these human experiences into a context appropriate to his own divine understanding of things. To be specific, the divine Word experienced as his own both the physical pain and the mental anguish of the crucifixion. Yet he was not overwhelmed by the experience, since in his divine self-awareness he recognized the value of such acute suffering for the redemption of the world.[8]

Turning now to Jesus in his human consciousness, it seems safe to say, as noted above, that he felt himself from his earliest years as an adult human being, perhaps even in childhood, under the influence of a Higher Power, prompting him to choose this rather than that, to act one way rather than another way. With his human will, of course, Jesus was in principle free to not obey the bidding of this inner voice and to follow instead self-centered cravings arising out of his subconscious. But presumably he never yielded to these inner cravings, at least not in a serious way, and thus paradoxically became freer and freer to obey the voice of the better self within him. That is, far from being less free as a result of being a divine

---

7. Ibid., 54. See also Karl Rahner, *Foundations of Christian Faith: An Introduction to the Idea of Christianity*, trans. William V. Dych (New York: Crossroad, 1978), 202, where he notes that Jesus is, like us, the recipient of divine grace but that, unlike us, he is also the privileged offer of grace to humanity from the Father.

8. Joseph A. Bracken, *God: Three Who Are One* (Collegeville, MN: Liturgical Press, 2008), 64; also Paul S. Fiddes, *The Creative Suffering of God* (Oxford: Clarendon Press, 1988), 62.

person as well as a full human being, Jesus was a great deal freer than the people around him. Whereas they were clearly influenced by guilt, fear, and anxiety as a result of past self-centered decisions, Jesus was liberated from all these compulsive behavior patterns by reason of his unswerving fidelity to the inner voice, which he came to see as the will of the Father, and the deeply rooted desire of his own better self (as he realized after the resurrection, the divine Word living within him).[9]

This being said, let us now move on to a process-oriented understanding of the doctrine of redemption. How did Jesus as the incarnate Word of God save his fellow human beings, redeem them from their fallen human condition as a result of original sin? Here we might first ask ourselves what is meant by the doctrine of original sin. With respect to the story of Adam and Eve and their fall into the power of Satan as a result of their ill-fated decision to disobey God's command in the Garden of Eden, many biblical scholars see it as symbolic rather than historical in character. It was meant to convey a religious truth about the perennial weakness of human nature in the face of temptation and the consequent need for God's grace rather than to be a strictly historical account of how sin and death first came into the world of creation.[10] In this way, one does not have to take issue with all the scientific evidence about the extended age of this world and the presence of suffering and death in the natural world long before human beings arrived on the scene as part of cosmic evolution. But if we assume that the first chapters of the book of Genesis are indeed symbolic rather than historical in character, what is the connection of these chapters with the doctrine of redemption?

In the history of theology, there have been several well-known theories explaining how Jesus redeemed us from our sins. In the early centuries of the church, the so-called ransom theory seemed to be favored by most theologians. That is, Jesus by his passion and death paid the price for our sins. But to whom was this ransom paid? If the ransom was paid to the devil, then the devil seems to have a power over human beings rivaling that of God the Father.[11] Likewise, if Christ suffered and died by way of expiation or sacrifice for our sins, then the connection of Christ's sacrificial death on the cross with the ritual sacrifices of animals in the temple under the Hebrew dispensation awakens misgivings about the relation between

---

9. Bracken, *The Triune Symbol*, 70.

10. See Jerry D. Korsmeyer, *Evolution and Eden: Balancing Original Sin and Contemporary Science* (New York: Paulist Press, 1998), 45–70.

11. See Louis Richard, *The Mystery of the Redemption*, trans. Joseph Horn (Baltimore, MD: Helicon, 1965), 149–56.

divine justice and divine mercy. Christ clearly represents to us the divine mercy, but what is to be said about divine justice in the person of God the Father?[12]

Then in the early Middle Ages, Anselm of Canterbury in his treatise *Cur Deus Homo* proposed that only someone who was both divine and human at the same time could render due "satisfaction" to God the Father for the offense given to the divine majesty by the sins of the human race.[13] A little later, Peter Abelard proposed that the redemptive value of Christ's death on the cross was that he revealed to us God's love and thereby enabled us to break free from our attachment to sin.[14] Abelard's theory, however, was hotly contested by Bernard of Clairvaux, who revived the patristic doctrine of the devil possessing certain rights over humanity; from psychological captivity to the devil, human beings can be released only by Jesus' sacrificial death on the cross.[15] Finally, in the late Middle Ages, Thomas Aquinas and John Duns Scotus took rival positions on the basic reason for the incarnation. For Aquinas, God became human simply as a remedy for human sinfulness. For Scotus, God became human to share the divine life with us, quite apart from the fact of human sinfulness. In this way, Aquinas sided more with the patristic and early medieval tradition, and Scotus sided with the more independent thinking of Peter Abelard.[16]

Something, of course, can be said for both sides of this argument. On the one hand, in Jesus' life, death, and resurrection, something objective must have happened in terms of the God-world relationship. If one of the three divine persons takes on a human nature and lives a fully human life, then salvation history, the history of our human relation to God, can never be the same again. But on the other hand, if human beings have to respond favorably to this cosmic event for it to have any significance for them as individuals, then Abelard and Scotus likewise have a point. With his life and message, Jesus was evidently setting forth a new way of life for human beings, which they might or might not accept. In either case, Jesus' life and message represented a challenge to their previous way of thinking and behaving, something which demanded a conversion of mind and heart from the people who heard what he had to say. One neglected this challenge at peril to oneself and one's future. Furthermore, Jesus evidently realized that he could not succeed in his life's work without assistance

---

12. Ibid., 155–68.
13. Ibid., 175–83.
14. Ibid., 184.
15. Ibid., 185–86.
16. Ibid., 200–205.

from other human beings. From the beginning of his ministry, therefore, he gathered a band of followers who were distinct from the other people who crowded around him to hear what he had to say and to be cured of their various afflictions by his healing touch.[17]

One can prescind here from the controversial question of whether Jesus wanted to reform Judaism or to establish another religion, which later came to be known as Christianity. The important point for our purposes here is that, in preaching the nearness of the kingdom of God, Jesus knew even then that he needed help from others, and he seemed also to realize that his small band of disciples would be the start at least of a reform movement, if not an institutional church, that would change the world by changing the way people related to God and to one another. Salvation or the coming of the kingdom of God, in other words, had for Jesus both an objective and a subjective dimension. He realized that it would not be enough for individual human beings to undergo a personal conversion of mind and heart as a result of his own life and message. These same individuals must then feel their affinity with one another as a result of that conversion to a new way of life and want to organize into a community so as to better share that life with one another and to spread the Good News of salvation to their neighbors as opportunities presented themselves. In the final analysis, then, Jesus with his life and message must have realized that he was starting something new that would last well beyond his earthly life. From this small band of disciples, a new corporate reality, something like a church or worshiping assembly, would inevitably arise.

All of this fits nicely with my revision of the relation between the One and the Many whereby the Many, subjects of experience in dynamic interrelation, coconstitute the One, a structured field of activity that outlasts its original component parts or members. As already mentioned, Jesus during his earthly life had a personal field of activity within the overall field of activity proper to the Israelite people of his day. But on a feeling level, he likewise participated in the much broader field of activity proper to himself as the divine Word within the Trinity. In the course of his public ministry, Jesus consciously established a new corporate field of activity to be shared with a group of chosen disciples, which he referred to as the kingdom of God. Thereby he was adding to the structure of the already existing divine-human field of activity that the three divine persons shared with all their creatures as a result of the initial act of creation, the beginning of the cosmic process. After his death and resurrection, what Jesus called

---

17. Joseph A. Bracken, *Christianity and Process Thought: Spirituality for a Changing World* (Philadelphia, PA: Templeton Foundation Press, 2006), 47–49.

the kingdom of God became focused in the church, the network of small communities of people around the Mediterranean world who consciously tried to live as Jesus did and who regularly celebrated the Eucharist as a symbolic expression of their union with him.

Still another feature of this process-oriented understanding of redemption, however, has to do with good and evil; promoting good and eliminating evil in this world is a necessary consequence of trying to live as Jesus did in preaching the kingdom of God. For that same reason, good and evil should be seen in the first place as collective realities, "powers and principalities," as Paul called them in his epistles.[18] Individuals, to be sure, make good and bad decisions. But the long-range effects of those same decisions contribute to a collective reality bigger than the individuals and the good and evil that they here and now bring about. In this sense, redemption should be understood as mobilizing the collective power of good against an ever-present and dangerous collective power of evil. The kingdom of God is then the stage on which the cosmic forces of good and evil engage in mortal combat. As I see it, Jesus himself recognized full well the character of this struggle, and for this reason alone he performed a large number of exorcisms during his public ministry.

In addition, if one accepts the field-oriented approach to Whitehead's notion of society, then one has a plausible explanation for why good and evil are necessarily collective or socially organized realities. For while individual actual entities at any given moment make particular decisions in terms of their own self-constitution, the impact of those decisions is always subtly to reinforce or in some modest way to modify the structure proper to the society or ongoing field of activity to which they belong.[19] Individual decisions, accordingly, always have social consequences, however inconsequential these consequences may seem to be at the time. Furthermore, a structure, once set in place by the collective decisions of an original group of actual entities, is not easily changed by later groups of actual entities. The longer the structure stays basically the same, the harder it will be for subsequent sets of actual entities to revise the structure that they have inherited.

---

18. Ibid., 42–44. Cf. also Walter Wink's celebrated trilogy: *Naming the Powers: The Language of Power in the New Testament* (Philadelphia, PA: Fortress Press, 1984); *Unmasking the Powers: The Invisible Forces that Determine Human Existence* (Philadelphia, PA: Fortress Press, 1986); *Engaging the Powers: Discernment and Resistance in a World of Domination* (Minneapolis, MN: Fortress Press, 1992).

19. Bracken, *Christianity and Process Thought*, 16–17.

There are, of course, distinct advantages in maintaining a continuity of structure within nature. We do not expect reality to be dramatically different from one moment to the next in our customary dealings with people and things. But as human beings quickly find out in trying to alter what they have come to regard as a negative behavior pattern in their own personal lives, the force of habit is very strong. We often have to be content with incremental changes in our behavior before a new, more positive habit takes shape and is then confirmed. Habitual patterns of group behavior, of course, are even harder to change since the majority of individuals in the group first have to recognize the disordered character of their group behavior and then have to make up their minds to behave differently in the future. Furthermore, group behavior patterns tend to get institutionalized in one way or another; that is, they are given the force of statutory law or long-standing custom.[20] One has only to recall the struggle of women and blacks for their civil rights here in the United States during the last century to recognize how a privileged group in society (males vs. females, whites vs. blacks) used the legal system to keep themselves in power and for a long time felt perfectly justified in doing so.

To return then to the topic of original sin and how it is to be conceived within this evolutionary context, we may say that original sin has far less to do with the sin of our legendary first parents than with the way in which all of us are heavily influenced in our individual behavior by the prevailing customs and laws of the communities into which we were born and in which we grew up.[21] We simply accepted without further reflection what our parents and other relatives, what our teachers in school, and in due time what our peers told us is the way things are and the way they should remain. Naturally, if we grew up in the right kind of family and/or community, we thereby picked up instinctively the right behavior patterns. If, on the contrary, we grew up in the wrong kind of family and/or community, we ended up adopting in unthinking fashion potentially destructive behavior patterns. But all human beings in one way or another have been negatively affected by the communities in which they grew up and, to some extent, in which they still live. For no human community is ever totally free of contamination by patterns of thought and behavior that are more destructive than beneficial in their effects on community members and many others as well.[22] In this sense, just as in the classical understanding of original sin stemming from the sin of Adam and Eve,

---

20. Ibid., 57–58.
21. Ibid., 44.
22. Ibid., 41–42.

no human being apart from Jesus and his mother Mary is exempt from the stain of original sin. Our personal sins in no small measure have their origin in the way we were brought up and in the way we subconsciously continue to live.

Moreover, one does not have to look far to see why this is so. As Plato argued centuries ago, communities tend to be individuals "writ large."[23] They are normally no better or no worse than the individuals who make them up. In fact, sometimes, as Reinhold Niebuhr claimed in *The Nature and Destiny of Man* years ago, the group is, or at least can be, "more arrogant, hypocritical, self-centered and more ruthless in the pursuit of its ends than the individual."[24] Especially when the community finds itself in direct competition with other communities for the attainment of certain goods and values, its members feel justified, out of loyalty to the community itself, to use whatever means might be necessary to guarantee its survival and continued well-being. At the same time, of course, communities can in a real sense be morally superior to the individuals who make them up, consciously more idealistic in their goals and values than what individuals simply by themselves normally strive to attain. We will see this more in a later chapter when I set forth a process-oriented approach to church and sacraments.

For the moment, however, we can safely say that both individuals and the groups to which they belong "sin" simply by becoming preoccupied with themselves and their own individual well-being. Thereby, of course, the members of the group lose a sense of contact with bigger human communities on the national and international level and with the wide world of nature all around them, to which they likewise belong and for whose survival and continued well-being they likewise have some responsibility. This is not to claim, of course, that the instinct of immediate self-preservation is itself sinful. It is, after all, a survival mechanism, something without which we human beings, both as individuals and as members of the human family, would not be here today. But like the classical doctrine of concupiscence, it is readily overextended and thereby abused. Niebuhr's point in the above citation was that individuals sometimes feel themselves freed from normal moral restraints when they are persuaded that the well-being or even the survival of the group is at stake. This is, of course, misguided idealism

---

23. See W. T. Jones, *A History of Western Philosophy*, 2nd ed. (New York: Harcourt, Brace and World, 1969), I, 167–70.

24. Reinhold Niebuhr, *The Nature and Destiny of Man: A Christian Interpretation*, vol. 1 (New York: Scribner's, 1941), 208.

since communities, like individuals, do not ultimately exist for themselves but for the sake of an even greater social reality, whatever that might be.

This same point, moreover, is why in my judgment Jesus in the Synoptic Gospels did not preach the coming of the church but the coming of the kingdom of God.[25] The two concepts were presumably in Jesus' mind, and certainly are in our minds today, closely related. But they are not identical. The church is a divinely chosen instrument for the advent of the kingdom of God. Hence, as the Second Vatican Council made clear in *Nostra Aetate* and other documents, explicit membership in the church is not indispensable for salvation.[26] One can presumably be saved, can gain admission to the kingdom of God both here and hereafter, by belonging to another Christian denomination than Roman Catholicism, by being a conscientious member of a non-Christian religion, or even by acting in good conscience as a nonbeliever, one without adherence to any religious institution. Christians, to be sure, believe that all human beings without exception are saved by the grace of Christ. But even so, one is drawn back to the fact that in the Synoptic Gospels, Jesus preached the coming of the kingdom of God and left to his followers the task of setting up the institutional church after Pentecost under the inspiration of the Holy Spirit.[27]

To sum up this chapter, Jesus the Christ in his life on earth was a divine person with both a divine and a human nature. That is, in terms of the process-oriented perspective laid out earlier, Jesus lived and was active in two quite distinct fields of activity that nevertheless significantly overlapped so as to create a common field of activity within which he could be both divine and human at the same time. In his divine nature, Jesus certainly realized better than he did in his human nature what was really going on. But even in his human nature, Jesus knew that he had a special relation to God, his heavenly Father, and that he was in a special sense God's Word, God's emissary to the people around him.

Jesus, accordingly, was sinless in his behavior. His sinlessness exhibited itself as an extraordinary awareness of and care for the needs of others even at considerable cost to himself. This accords well, of course, with his status as a divine-human person. The self-giving love characteristic of the

---

25. Bracken, *Christianity and Process Thought*, 48–49.

26. *The Documents of Vatican II*, ed. Walter M. Abbott, SJ (New York: Guild Press, 1966): *Dei Verbum*, nos. 2–6; *Gaudium et Spes*, nos. 1–3; *Nostra Aetate*, nos. 1–5.

27. Cf. on this point Bernard Prusak, *The Church Unfinished: Ecclesiology through the Centuries* (New York: Paulist Press, 2004), 9, 56–69; likewise, Richard Gaillardetz, *Ecclesiology for a Global Church: A People Called and Sent* (Maryknoll, NY: Orbis Books, 2008), 18–19. Both authors maintain in different ways that Jesus initiated a movement that in due time became the institutional church.

relations of the three divine persons to one another within the divine life found full expression in the way that Jesus dealt with his contemporaries. Accordingly, much as first Peter Abelard and then John Duns Scotus insisted, the deeper reason for the incarnation seems to have been not simply to make satisfaction for human sinfulness but to share the divine life with the human race and in this way to inaugurate a new era in human history where self-giving love would have the final word. Justice must still prevail, but in the end mercy (both divine and human) triumphs over justice.

Finally, all this takes place within a broadly communitarian and processive context. The kingdom is an "already-but-not-yet" phenomenon, something already here in our midst and yet still not present in its full actuality.[28] We can even now experience the effects of this new divine life within us and among us, at least in our better moments, but we also sense that for us as individuals and for the human race as such, the full reality of this new divine life is yet to come. Likewise, we must never forget that this world of ours is dominated by cosmic "powers and principalities" in St. Paul's sense of the terms. Good and evil are grounded in the decisions of individual subjects of experience. But these decisions have social consequences in terms of the prevailing structure of the various fields of activity within which we live. Each of us in our individual decisions and in the determinations of the communities to which we belong contributes to both the collective power of good and the collective power of evil. The important thing is that we recognize the seriousness of this cosmic struggle and align ourselves with Jesus in giving new impetus and direction to the collective power of good.

---

28. See Hans Schwarz, *Eschatology* (Grand Rapids, MI: Eerdmans, 2000), 135–51.

10

A New Look at the Resurrection of the Body

In a recent book on the relation between modern science and traditional religious beliefs, English philosopher/theologian Keith Ward takes note of the fact that in the view of some scientists human thoughts and memories may someday be "downloaded" into supercomputers that would far outlast the possibility of physical human survival on this earth.[1] In this way human beings could achieve a type of natural (as opposed to supernaturally caused) immortality.[2] Ward then comments: "What computer analogies really show is not that we are just machines. They show that we—the very same conscious intelligent and responsible agents with a rich inner life of memories, hopes, and fears—could possibly be reembodied in very different forms."[3] This is an intriguing idea but one that in my judgment requires further analysis. At issue, for example, is what is meant by a body or, for that matter, any sensibly perceptual reality or "thing."

Ward himself raises that question elsewhere in his book: "We might not want to call the colors, sounds, and smells we all subjectively experience 'entities,' as though they were objects in a shadow world paralleling the physical universe. But they are appearances that are different from the world that is the causal object of their appearing. They are appearances of the world to consciousness, and consciousness itself is the way the world appears in immediate experience."[4] So if a tree in midsummer is itself not green but only looks green because it is perceived as such by a human

---

1. Keith Ward, *The Big Questions in Science and Religion* (West Conshohocken, PA: Templeton Press, 2008), 152.
2. See Frank Tipler, *The Physics of Immortality: Modern Cosmology, God, and the Resurrection of the Dead* (New York: Doubleday, 1994), 1–15.
3. Ward, *Big Questions*, 152.
4. Ibid., 157.

brain with a very specific visual perceptual system, then what is a physical body apart from our normal human perception of it as something solid and enduring? If one is to claim that the human mind or soul can move from one form of physical embodiment to another, then the question remains: What is a body, in particular a human body, so that a human being could be equally "at home" in a new "resurrected" body as well as in a normal physical body? Likewise, for a Christian who concomitantly believes in the reality of the Mystical Body of Christ as somehow even now encompassing the whole of creation (cf. Eph 1:3-10; Col 1:12-20), one must inevitably ask about the relation between one's physical body here and now, one's resurrected body after death, and the bodily reality of the risen Christ.

Given Ward's generally favorable treatment of the process metaphysics of Alfred North Whitehead,[5] I offer in this chapter an explanation of "body" as what Whitehead calls a "structured society" or society composed of subsocieties of actual entities, momentary self-constituting subjects of experience.[6] But just as elsewhere in this book, I revise Whitehead's own understanding of "societies" so as to consider them as structured fields of activity for their constituent actual entities rather than as simply collections of genetically interrelated actual entities at any given moment. From this starting point, I offer my own explanation for Christian belief in the resurrection of the body. This chapter, accordingly, will be divided into two parts. First, I will briefly summarize my argument for rethinking Whitehead's category of society as a structured field of activity for its constituent actual entities at any given moment. Then, I will dialogue with Keith Ward on how best to explain life after death as full incorporation into the kingdom of God or the Mystical Body of Christ.

Whiteheadian Societies as Structured Fields of Activity

For Whitehead, a society is clearly more than a random aggregate of actual entities, momentary self-constituting subjects of experience. They are genetically related to one another in virtue of common inheritance of a "common element of form" or "defining characteristic." Yet given Whitehead's acceptance of the strictly analytical approach to physical reality characteristic of the natural sciences of his day whereby physical

---

5. Ibid., 129–32, 251–53; see also Keith Ward, *Pascal's Fire: Scientific Faith and Religious Understanding* (Oxford: One World, 2006), 162–64.

6. Alfred North Whitehead, *Process and Reality: An Essay in Cosmology*, corrected edition, ed. David Ray Griffin and Donald W. Sherburne (New York: Free Press, 1978), 34, 99.

bodies and all other material "things" are ultimately composed of atoms (in Whitehead's case, spiritual atoms or momentary subjects of experience[7]), one can only conclude that Whitehead still thinks of societies as in some sense aggregates of ultimate components. As he comments elsewhere in *Process and Reality*, "All the life in the body is the life of the individual cells. There are thus millions upon millions of centres of life in each animal body."[8] So is a Whiteheadian society then only an aggregate of component parts, a whole made up of the sum of its parts, however intricately interrelated, or is it something else, namely, a specifically social reality different from and other than its component parts, yet somehow dependent upon those same parts for its own existence?

My response to Whitehead's apparent conceptual dilemma here is to affirm the second alternative mentioned above. A Whiteheadian society is a specifically social reality, an ontological whole different from but still dependent upon its component parts for its own existence. It is, in other words, an entity originally emergent out of the dynamic interplay of component parts and yet, once established in being, possessing a form of existence proper to itself as a specifically social or corporate reality. In this respect, it is like an Aristotelian substance governed by an unchanging substantial form.[9] Yet because of its ongoing interdependence with its component parts in their dynamic interrelation, a Whiteheadian society must also be capable of transformation or evolution in terms of its defining characteristic. To account for both these factors in equal measure, in my judgment, a society is best understood as a structured field of activity for its constituent actual entities. For on the one hand, it is then subject to change in its defining characteristic, given its dependence upon the way its ever-changing components dynamically relate to one another. On the other hand, it clearly exercises a constraint on its constituent actual entities when as momentary self-constituting subjects of experience they come into being and then seek to perpetuate themselves in terms of their influence on the shape or pattern of existence for their successors in the same society. Understood as an already existing structured field of activity for its constituent actual entities of the moment, therefore, a Whiteheadian

---

7. Ibid., 18: "Actual entities—also termed 'actual occasions'—are the final real things of which the world is made up. There is no going behind actual entities to find anything more real." See also ibid., 35: "The ultimate metaphysical truth is atomism. The creatures are atomic."

8. Ibid., 108.

9. Alfred North Whitehead, *Adventures of Ideas* (New York: The Free Press, 1967), 204.

society is the necessary ontological context or environment for those actual entities. Not in virtue of a fixed substantial form, as in the classical understanding of an Aristotelian substance, but through its informational content for each new set of constituent actual entities, a Whiteheadian society exercises a type of formal causality. Within an Aristotelian substance, to be sure, the substantial form is active; within a Whiteheadian society, the common element of form is passive—that is, it is available for "prehension" by each new set of self-constituting actual entities, but it is still quite real in its overall effectiveness.

Application to the Notion of the Resurrection of the Body

At this point we can return to Keith Ward's proposal that human beings after death in this life could possibly be reembodied in a "resurrection world," located somewhere else in this universe or in some other universe existing parallel to our own.[10] His point is not to claim that this will happen but only to say that contemporary cosmologists and other scientists cannot dismiss such a claim as impossible, given their current understanding of the laws of nature. Furthermore, a religious conception of the ultimate goal of the cosmos is necessarily far less concerned about the physical end of our universe—or indeed of any other universe existing before, during, or after our own universe—than about "the final flourishing of conscious life that is the divine purpose of the universe."[11] For in this way, "the sufferings of finite creatures will be transformed and fulfilled by a deeper and fuller conscious relationship to the Supreme Good, in which physical laws have always been grounded."[12] Keeping these specifically religious goals and values in mind, I now sketch how a revised understanding of a Whiteheadian society as a structured field of activity for its constituent actual entities can be used to justify a panentheistic understanding of the God-world relationship in which finite creatures come forth from God, exist for a while in partial independence of God, but in the end are incorporated into the fullness of the divine life. Furthermore, this full incorporation of finite creatures into the divine life can be seen as taking place at every moment within the cosmic process in a manner somewhat akin to Whitehead's understanding in *Process and Reality* of the workings of the divine consequent nature but with the added qualification of subjective as well as objective immortality for those same finite subjects

---

10. Ward, *Big Questions*, 158–61. See also *Pascal's Fire*, 233–58.
11. Ward, *Pascal's Fire*, 256.
12. Ibid., 255.

of experience. But in that case, creatures do not have to be transported from one universe to another, or at least from one part of our current universe to another part, in order to participate in the "resurrection world" spoken of by Ward. Resurrection is better presented as an awakening to the fullness of the world that one has continually experienced from birth but only dimly understood.

Yet Whitehead's own conception in *Process and Reality* of the God-world relationship is not panentheistic in inspiration with the world existing somehow in God and yet distinct from God. Rather, in Whitehead's view God is very much a part of the cosmic process: "God and the World are the contrasted opposites in terms of which Creativity achieves its supreme task of transforming disjoined multiplicity, with its diversities in opposition, into concrescent unity, with its diversities in contrast."[13] A panentheistic, process-oriented understanding of the God-world relationship can be found in the writings of Charles Hartshorne, John Cobb, and other Christian Whiteheadians. They conceive God as the "soul" of the world and the world as the "body" of God.[14] Although this approach to panentheism is clearly more in line with the biblical conception of God, it also has certain limitations. For as Sallie McFague and others have pointed out, God is still dependent upon the world for God's own raison d'être, and creatures—above all, human beings—are equivalently God's body parts.[15]

My own version of a process-oriented panentheism emphasizes the trinitarian reality of God.[16] That is, each of the divine persons is conceived as a personally ordered "society" of divine actual occasions; the three divine persons are nevertheless one God in virtue of being together a Whiteheadian structured society, a society of subsocieties of actual occasions on the divine level of existence and activity. Then, in line with my rethinking of the notion of society as a structured field of activity for its constituent actual entities, I stipulate that, while each of the divine persons as a personally ordered society of actual entities presides over a field of activity proper to its own existence and activity, these three individual

---

13. Whitehead, *Process and Reality*, 348.

14. See, for example, Charles Hartshorne, "The Compound Individual," in *Philosophical Essays for Alfred North Whitehead*, ed. F. S. C. Northrup (New York: Russell & Russell, 1936), 193–220.

15. See Sallie McFague, *Models of God: Theology for an Ecological Nuclear Age* (Philadelphia, PA: Fortress Press, 1987), 69–77. See also Joseph A. Bracken, *Christianity and Process Thought: Spirituality for a Changing World* (Philadelphia, PA: Templeton Press, 2006), 5–7.

16. See Bracken, *Christianity and Process Thought*, 7–13, 107–15, for a more extensive explanation of my process-oriented interpretation of the God-world relationship.

fields of divine activity are fully integrated into a collective field of activity proper to themselves as one God, a divine community. Finally, within this all-embracing divine field of activity proper to the three divine persons, the world of creation has slowly taken shape, beginning with the big bang fourteen billion years ago.[17]

In my judgment, this metaphysical scheme is better adjusted to classical Christian belief than either Whitehead's dialectical understanding in *Process and Reality* of the God-world relationship or the notion of panentheism presented by Charles Hartshorne, Sallie McFague, and others in terms of a soul-body metaphor. For as already noted, it presupposes classical Christian belief in the doctrine of the Trinity and allows for preservation of traditional Christian belief in *creatio ex nihilo* (creation from nothingness). That is, the three divine persons within this scheme clearly share a communitarian life, quite apart from involvement in the process of creation. As a result, the world of creation presumably came into being within the divine field of activity by a free decision on the part of the divine persons to share that communitarian life with creatures, above all, with human beings as likewise capable of rational thought and free decision. It is, of course, an understanding of *creatio ex nihilo* not as creation from absolute nothingness but as creation out of the energy resources proper to the divine persons in their ongoing collective field of existence and activity. It is in this sense perhaps better presented as *creatio ex Deo*, provided that *Deo* means not God as a personal being, but the nature of God, the vital source of the divine being or the divine energy field.[18] Finally, it also seems to correspond nicely to much of what Keith Ward offers as a scientifically as well as religiously grounded understanding of eternal life or, more specifically, the resurrection of the body from a Christian perspective.

As already noted, Ward suggests that human beings after the death of the body could be instantaneously transported by God either to some distant place in our universe or into another universe altogether, where they could enjoy a much closer union with God as the source of all being and as the final goal of the cosmic process. Yet such an intimate union with God "should be in some intelligible causal relationship to the life lived on earth. It should enable persons to see their earthly life in the wider context of the divine knowledge and experience of all things. It should enable people to recognize and come to terms with the evil that

---

17. I have explained this field-oriented approach to panentheism in many books and articles, beginning with *The Divine Matrix* (see below, n. 18).

18. See Joseph A. Bracken, *The Divine Matrix: Creativity as Link between East and West* (Maryknoll, NY: Orbis Books, 1995), 52–68.

they have done and perhaps to find some way of making amends for it. And it should enable persons to grow further in knowledge and love of the Supreme Good, in ways impossible on earth."[19] Within the parameters of the God-world relationship that I have just laid out, all these conditions would seem to be satisfied.

For if, as noted above, creation as a vast Whiteheadian structured society, that is, as a society composed of innumerable subsocieties of actual entities, comes into being and continues to exist within the divine field of activity coconstituted by the three divine persons in their ongoing dynamic interrelation, then heaven and earth are inseparable. That is, earth exists within heaven, the all-embracing divine field of activity that Scripture describes as the kingdom of God or the Mystical Body of Christ. Thus, we human beings and indeed all of creation do not have to be transported somewhere else to live in union with God. As St. Paul says in Acts 17:28, even now "we live and move and have our being" in God. Admittedly, because of the inherent limitations of life in the body, we do not see the bigger picture, but presumably at the moment of death we are freed from those bodily limitations and can see for the first time the meaning and value of our lives in terms of the enduring structure of the common field of activity that we share with the divine persons and all finite entities and to which we have contributed by the "decisions," conscious and unconscious, that we made from moment to moment during our lifetime.

Perhaps the best way to make this clear is to contrast it with Whitehead's notion of objective immortality in terms of the divine consequent nature, God's ongoing integration of everything that happens within the cosmic process from moment to moment. The notion of the divine consequent nature is Whitehead's way of avoiding what he considers to be the greatest evil in the world, namely, "perpetual perishing."[20] Since God is the sole nontemporal actual entity,[21] who "prehends" everything that comes into being and whose process of concrescence never ends, God can reconcile with one another all the apparently discordant events that take place at every moment in the cosmic process and give each event its place within the "completed whole": "The revolts of destructive evil, purely self-regarding, are dismissed into the triviality of merely individual facts; and yet the good they did achieve in individual joy, in individual sorrow, in the introduction of needed contrast, is yet saved by its relation to the

---

19. Ward, *Big Questions*, 161.
20. Whitehead, *Process and Reality*, 340.
21. Ibid., 46.

completed whole."[22] This is a wonderfully poetic statement of God's work in the world. But from a purely logical perspective, it is a category mistake. Whitehead is confusing God's subjective experience of the world with the world's objective reality.[23] God's subjective experience of the world, to be sure, is necessary for God to transmit divine "initial aims"[24] to the next set of concrescing actual entities within the cosmic process. But a brief look at the state of affairs in the world today should convince any reasonable person that the harmony in God's idealized experience of the world does not yet correspond to objective reality.

Given his presupposition that God alone survives the passage of time and that all other actual entities perish as soon as they come into being, Whitehead had to make God's experience of the world coincident with the world's objective reality from moment to moment. But if he had chosen to give societies an objective reality distinct from the moment-by-moment interplay of their constituent actual entities (e.g., as enduring structured fields of activity for successive sets of actual entities), then he would have been able to distinguish between God's subjective experience of the world at any given moment and the corresponding objective reality of the world as a complex structured society, a society of subsocieties of actual entities, which enjoys its own form of existence within the divine field of activity. In any case, given the hypothesis that human beings and all other finite entities by their moment-to-moment self-constituting decisions contribute to the ongoing structure of the common field of activity that they share with the divine persons, then one can argue that if by a special favor (grace) from the divine persons human beings and other finite entities with some degree of consciousness are able to survive the death of their bodies and thus to comprehend their place in the cosmic process and their relationship to the divine persons, they would be in heaven. That is, they would be enjoying the divine life free of the limitations of life in this world. Human beings, of course, would have to accept the full truth about themselves, given this comprehensive overview of their lives as part of a cosmic process much bigger than themselves. This, in turn, would require taking responsibility for their sinfulness, the evil as well as the good that they brought about by their past behavior. In line with the teaching of Jesus in the Gospels, the divine

---

22. Ibid., 346.

23. As I have argued in chapter five, Thomas Aquinas made a similar mistake in identifying God's essence and existence (see Thomas Aquinas, *Summa Theologica*, I, q. 3, a. 3). Existence is not the greatest perfection in the hierarchy of forms or essences. It simply denotes what is the case, an objective state of affairs.

24. Whitehead, *Process and Reality*, 244.

persons would presumably be ready to forgive humans if they openly acknowledged their wrongdoing. But this would demand an unaccustomed degree of humility from at least some human beings who, given their self-centered behavior in their earthly lives, might find it very difficult to admit their guilt before the divine persons and their fellow human beings in the afterlife. Hence, within this scenario for life after death, there is provision for a prolonged purgative state before full incorporation into eternal life, which would be roughly equivalent to Roman Catholic belief in purgatory, minus the flames and other forms of corporal punishment. There is even provision for a self-inflicted "hell" if one obstinately refuses to accept one's sinfulness and the offer of forgiveness from the divine persons and thus chooses to live in isolation from God and other human beings.[25]

So it would seem that this neo-Whiteheadian approach to life after death meets the conditions laid down by Ward in his book *The Big Questions*. To repeat the passage cited earlier, a Christian understanding of life after death means that "it should be in some intelligible causal relationship to the life lived on earth. It should enable persons to see their earthly life in the wider context of the divine knowledge and experience of all things. It should enable people to recognize and come to terms with the evil that they have done and perhaps to find some way of making amends for it. And it should enable persons to grow further in knowledge and love of the Supreme Good, in ways impossible on earth."[26] One could, of course, object that this notion of heaven on earth seems to reduce human beings and all other finite entities after death to ghosts inhabiting a spirit world invisible to those who still exist in space-time. Or is it the case that we modern human beings have become so mesmerized by the indubitable success of the natural sciences and their related technologies that we live in a "one-storey" universe, a world of human experience shrunken to the level of what can here and now be empirically perceived and mathematically measured? This is the argument of Huston Smith, the celebrated author of *The World Religions*, in a subsequent book titled *Forgotten Truth*.[27] Without trying to directly settle this highly controversial issue, I turn instead to analysis of what Whitehead says in *Process and Reality* about the dual-dimensional character of actual entities and ask what that

---

25. All this is laid out in much more detail in my book *Christianity and Process Thought*, 103–15.
26. Ward, Big Questions, 161.
27. Huston Smith, *Forgotten Truth: The Primordial Tradition* (New York: Harper & Row, 1976), 1–18; see also *The World's Religions* [originally, *The Religions of Man*] (New York: HarperCollins, 1991).

might mean for the basic topic of this chapter, namely, belief in the resurrection of the body.

An actual entity, says Whitehead, is constituted by two "poles," one mental or spiritual and the other physical or material.[28] This is what he evidently meant in describing an actual entity as a "subject-superject."[29] The actual entity as "subject" is a momentary self-determining subject of experience; as such, it is an immaterial reality that cannot be directly prehended by any other actual entity, not even by God. The actual entity as "superject," however, is a material reality that can be prehended by other actual entities, including God. But what is it that is thus prehended? Whitehead is not entirely clear on this matter. In *Process and Reality*, he notes that "in the philosophy of organism it is not 'substance' which is permanent, but 'form.'"[30] This would lead one to conclude that an actual entity as superject is only prehensible in terms of the form or structure of its own self-constitution as an objectified momentary subject of experience. Hence, it has no physicality or materiality beyond being an intelligible form or ministructure that, in terms of my own understanding of a society as a structured field of activity for its constituent actual entities, is woven into the overall structure or common element of form characteristic of the society to which it belongs. In other words, what we human beings perceive as macroscopic physical realities (organisms or inanimate things) are in fact enduring patterns of dynamic interrelation among immaterial self-constituting subjects of experience (actual entities). Organisms and things are only apparently "substances," independently existing material or purely physical realities. Considered apart from human perception, organisms and things are instead enduring complex patterns of dynamically interrelated energy-events. Presumably as a technique for survival in a hostile natural environment, human beings and other higher-order animal species with a brain and central nervous system have learned to simplify the dynamic energy-exchanges going on all around and inside themselves by reducing these psycho-physical events to apparently solid and enduring ("substantial") physical realities.

The logical consequence of this line of argument, of course, is that the state of a human being at the moment of death and resurrection (or indeed of any other finite entity after its demise as a physical reality perceptible to the senses) is not that of a ghost or disembodied reality. It still possesses a body insofar as it continues to inhabit a structured field of activity with

---

28. Whitehead, *Process and Reality*, 208–15.
29. Ibid., 27–28.
30. Ibid., 29.

a concrete historical pattern of existence and activity. But that pattern of bodily existence and activity peculiar to each of us as individuals has been also incorporated into the governing structure of even larger fields of activity proper to the communities and environments within which we have lived all these years. Likewise, the pattern of existence and activity proper to us as individual human beings has been progressively incorporated into the Mystical Body of Christ or the kingdom of God, namely, that structured field of activity common to both the divine persons and all their creatures (from the simplest to the most complex) for their ongoing dynamic interrelation. But here one may object that the body of the risen Jesus appeared to be a normal human body, not a complex pattern of interrelated energy-events. In response, I would argue that the disciples in seeing the risen Jesus were just as limited in their powers of perception as they were in observing him during his earthly life. That is, they saw him as a tangible material reality, not as a "spiritual" body freed from the limitations of space and time. Thus, unlike images of the risen Lord as somehow the same person as before his death on the cross, Jesus was not miraculously restored to normal human life by God the Father on Easter Sunday. His risen body is now fully incorporated into the field of activity proper to his divinity; yet it retains its own identity as a finite field of activity so that Jesus can manifest himself to his disciples as seemingly the same person as before his death. For only in this way can Jesus be present to his followers wherever they are in this world; only in this way can he be present under the forms of bread and wine at Eucharistic celebrations simultaneously taking place around the world.

But, one may ask, if Jesus rose into the fullness of his divinity at the moment of his death on the cross, what happened to his corpse or physical remains afterward? What should be said about the empty tomb on Easter Sunday morning? To answer this question, we are once again drawn back to consideration of what we mean by a physical body. Is it a tangible material reality existing in its own right, or is it a moment-by-moment intricate pattern of dynamically interrelated energy-events? If it is the former, one logically has to claim that Jesus on Easter Sunday simply reanimated his physical body and somehow turned it into a spiritual body that will never die nor suffer any of the limitations of earthly life in the body. But is this possible? Would the inevitable constraints of life in an earthly body make impossible the new freedom of life in a spiritual body? There is no way to directly answer that question. But if a physical body is not a material reality but rather an intricate pattern of dynamically interrelated energy-events, as noted above, one can claim that this pattern of dynamically interrelated energy-events broke down much faster than is normal with a dead body, so that by Easter Sunday morning the earthly body of Jesus no longer existed in a recognizable form. When the followers of Jesus arrived at the tomb on Easter Sunday morning and looked inside, the earthly body of Jesus was no longer there. The energy-potential that sustained his earthly body had been reincorporated into the energy-pool / structured field of activity proper to the earth as an ecological system, the created universe, and the divine field of activity proper to the three divine persons. Certainly Jesus was there in virtue of his new status as the risen Lord, but he was not visible to those who came to the tomb until he chose to reveal himself to them in the unexpected way described in the Gospel accounts. In brief, then, I argue that to live the new state proper to a risen body, Jesus as a divine person with a human nature had to allow the state proper to his earthly body to cease to exist. He no longer inhabits a material body but a risen or spiritual body[31] with potentialities for existence and activity that would be absolutely impossible for life in a reanimated material body.

One more point of comparison between Ward's approach to life after death and my own proposal using a revised understanding of Whiteheadian societies should be mentioned before concluding this chapter. In his book *Pascal's Fire*, Ward offers the following thought-provoking remark about scenarios for the end of the world, both those found in the Bible and in other sources:

> The history of the cosmos, as I have presented it, is an expression of ultimate mind, bringing into being by a long evolutionary process other minds, physically based and physically generated, that can share, in an appropriate way, the self-awareness and creativity of their originator. But within the cosmos such worthwhile states may exist maximally as the mid-point of the life of the cosmos, in its mid-life maturity, and not at its end.[32]

So the apocalyptic passages in both the Hebrew and Christian Bible that describe the Last Judgment and the end of the world as simultaneously occurring should not be taken literally. At least in the New Testament, they may well be a mixture of historical events already experienced by Christians who lived in or near Jerusalem at the time of its destruction in 70 CE and Jesus' own predictions about the end of life on this earth.

In line with my own neo-Whiteheadian understanding of eternal life in terms of full incorporation into the proposed divine field of activity,

---

31. See 1 Cor 15:44: "It [the human body] is sown a natural body; it is raised a spiritual body. If there is a natural body, there is also a spiritual one."

32. Ward, *Pascal's Fire*, 242.

however, it should also be possible to propose that the Last Judgment takes place for human beings and all other finite entities at the moment that their physical existence on this earth comes to an end. Here I am strongly influenced by another Whiteheadian scholar, Marjorie Suchocki in her book *The End of Evil*. Therein she proposes that God in virtue of the divine consequent nature prehends each finite actual entity not simply in its objective reality as a superject but likewise in its reality as a realized subject of experience, "enjoying" what it has become at the end of its process of self-constitution.[33] This is a highly creative way to justify subjective immortality, at least for human beings, within a basically Whiteheadian frame of reference. But in my judgment it suffers from the same tendency to logical atomism as Whitehead's own metaphysical scheme. Pace Whitehead and Suchocki, what are saved for eternal life with God are not individual actual entities but the societies to which they belong. For this to happen, at least in the case of human beings, all that is needed is for the final actual entities constituting a human being as a complex structured society of actual entities in space and time to be granted full participation in the divine life by way of divine favor or special grace. All the preceding moments of experience (actual entities) proper to that human being over his or her lifetime will have naturally ceased to exist and will be present to that final set of actual entities only insofar as they in some modest way added to the ongoing structure of the field of activity proper to that person at the moment of death.

Then, at the moment of death, the nexus of actual entities constituting the soul of that human being will be granted by divine grace an enormously expanded sense of self-possession or self-awareness. That person will see for the first time the full pattern of his or her past life as inscribed within the ongoing structure of the field of activity common to itself, to other creatures, and to the divine persons. In this way a human being will have a privileged opportunity, as noted above, either to humbly accept or to arrogantly reject the truth of his or her entire past existence. Thus there will be upon entrance into eternal life a moment of judgment; but it will be a strictly internal or self-imposed judgment, not an external judgment conferred on an individual by an angry God.

So the Last Judgment could conceivably take place at the moment each of us dies or ceases to exist in the space-time continuum proper to earthly life. One now exists in the "duration" or time-frame of God, which theologians have traditionally called eternity. Like the divine persons, one

---

33. Marjorie Hewitt Suchocki, *The End of Evil: A Process Eschatology* (Albany, NY: State University of New York Press, 1985), 81–96.

is presumably simultaneously present to all successive moments of time in the cosmic process.[34] Thus, one's particular judgment at the moment of death coincides with the Last Judgment at the end of the world in the scriptural view. Furthermore, not only the mind or soul of a human being as a privileged nexus of actual entities within the structured society proper to a human being is saved for eternal life, but also the various subsocieties corresponding to the body of that human being are also saved. Since I am a dynamic union of soul and body in this life, my body will also share in eternal life. Like the risen body of Jesus, however, my earthly body will undergo a change of state, become a spiritual body, so as to participate in eternal life. Finally, if the physical body of a human being can be in this way resurrected for eternal life, then in principle every finite reality within the world of creation can achieve at the end of its existence on earth some limited share in the fullness of the divine life, albeit without undergoing anything like a Last Judgment, which is only proper to human beings as responsible creatures made in the image and likeness of God.[35]

I conclude this essay with Ward's comments on whether the universe has an ultimate goal or purpose:

> Has the universe a goal? Science does not usually ask that question. But for many, not all, cosmologists, it may have, and if so, the goal has to do with the increase of knowledge, freedom, and intelligent life. Will the goal be realized? Not literally forever, in this universe, it seems. But it may be realized for a while, or it may be realized more fully if there are universes of an appropriate sort beyond this. . . . For much religious thought, the goal will be realized, in another universe, which may be a transfigured form of this one.[36]

The point of this chapter has been to question whether the goal of increased "knowledge, freedom, and intelligent life" will require another universe than the one in which we already live. Perhaps our universe is already being transformed in an invisible way at every moment without our realizing it. I say "in an invisible way" simply to make clear what Ward himself contends, that the full reality of the world is invariably more

---

34. See Joseph A. Bracken, *Subjectivity, Objectivity and Intersubjectivity: A New Paradigm for Religion and Science* (West Conshohocken, PA: Templeton Press, 2009), 168–78, where I discuss issues related to time and eternity in both religion and science.

35. For a more extended treatment of the Last Judgment and the end of the world in the context of my neo-Whiteheadian scheme, see my book *Christianity and Process Thought*, 103–15.

36. Ward, *Big Questions*, 58.

than what human beings here and now perceive.[37] As noted above, in the interests of survival and reproduction, human beings and other animal species with a central nervous system and a brain have found it necessary to simplify the real world of complex, interrelated energy-events by reducing it to a world of appearances in which we and other finite entities seemingly exist apart from one another as separate physical realities. The cessation of existence of what we here and now call our bodies removes the limitations of perception thereby imposed and allows us to participate more fully in the real world of dynamically interrelated energy-events, which Scripture calls the kingdom of God or in other places the Mystical Body of Christ.

This is not an outrageous claim without a shred of empirical evidence. Quantum mechanics has already made clear to us the reality of an invisible world of dynamically interrelated energy-events all around us and within us. But if this is so, perhaps we should focus our speculation about the world to come on possibilities already to be found within nature in this world but still only imperfectly understood. In other words, perhaps we should withhold judgment on proposals from secular cosmologists about the various ways in which human beings may someday achieve salvation, unending life, elsewhere in the current universe or in some other universe through advanced scientific technology. Yet even with this reservation about Ward's reflections on life after death, I fully endorse his more important claim that the possibility of eternal life for human beings and the other finite creatures of this world must be founded on "a firm belief in the reality of God, or spiritual being."[38] God alone possesses by nature unending life. If we human beings and other finite creatures achieve salvation, it will only be by participation in the divine life as the free gift of a loving God. Here and now, natural science can assist us to make such a religious belief more plausible, but no theory of natural science can substitute for that same belief.

---

37. Ibid., 157.
38. Ibid., 58.

11

Church and Sacraments from a Process Perspective[1]

While classical Thomistic metaphysics and transcendental Thomism have different starting points, the one in cosmology or metaphysics, the other in the theory of knowledge or epistemology, both implicitly affirm one and the same understanding of the relation between the One and the Many. That is, both presuppose the priority of the nonempirical One over the empirical Many, as initially proposed by Plato. This classical Platonic approach to the problem of the One and the Many has, moreover, entered deeply into the institutional life of the Roman Catholic Church. The ontological priority of the One over the Many is reflected in the priority of the Pope as the Vicar of Christ over his brother bishops, the priority of the bishop over the priests in his diocese, and the priority of the pastor over the parishioners in the local parish. For that matter, this same thought pattern is reflected in the conscious or unconscious attitude of many Roman Catholics to ecumenical dialogue with members of the other Christian denominations and to interreligious dialogue with representatives of the other major world religions. As they see it, the Roman Catholic Church should be the basic point of reference, or the One, in dealing with other Christian denominations and the non-Christian world religions since these other religious bodies are collectively the derivative reality of the Many. Roman Catholics, they claim, can certainly learn from their dialogue-partners, but

---

1. A somewhat lengthier version of this chapter can be found in a book of essays by different authors on the pertinence of Whitehead's metaphysics for interreligious dialogue to be published by Wipf and Stock (Eugene, OR) under the editorship of John B. Cobb, Jr.

in the end the truth to be gained in and through the dialogue will find its full expression only in current Roman Catholic belief and practice.

Since the second half of the twentieth century, however, a new mind-set seems to have taken hold, at least among some Roman Catholics within the institutional church. In many of the documents of the Second Vatican Council, for example, there was a totally unexpected focus on reform and renewal of church life "from the bottom up," so to speak. Furthermore, the enthusiastic reception of those same reform-minded documents in certain quarters was clear testimony to a felt need among many Catholics for change in traditional belief and practice. From a philosophical perspective, what seems to have happened here was an awakening to the inevitable limitations of the classical paradigm for the relation between the One and the Many. More attention should be paid to the reality of the empirical Many (the People of God) and to the ultimate dependence of the One (the institutional church with its established beliefs and practices) on the thinking and behavior of the laity, the people in the pews every weekend.

Within this post–Vatican II ferment among liberally oriented Roman Catholics, Bernard Lee, SM, published *The Becoming of the Church*, a book on church and sacraments inspired by the thought of Whitehead and some early Whiteheadian process theologians, notably John Cobb.[2] In what follows, I will first summarize the contents of this important work so as to make clear how Whitehead's philosophy can have a significant impact on the Christian understanding of church and sacraments. Afterward, I will offer some comments on how Lee's book could have even more influence on the thinking not just of Roman Catholics but of other Christians as well if it were a little more nuanced in its presentation. For Lee's dramatic shift in focus from the One to the Many (i.e., from official belief and practice within the Church to felt beliefs and actual practices of people in the pews) was perhaps too big a change for even quite liberally oriented Roman Catholics at that time to assimilate. What might be needed, then, for broader acceptance of the tenets of process theology among Roman Catholics could well be a middle-ground position whereby top-down causality in the form of traditional church authority, the teaching of the Pope and bishops, can be nicely reconciled with a new awareness of bottom-up activity from within the church, that is, the ongoing response of the laity to current church teaching. This balancing act, of course, will require some rethinking of Whitehead's metaphysical category of "society," as I have already made clear in previous chapters of this book.

---

2. Bernard Lee, SM, *The Becoming of the Church: A Process Theology of the Structure of Christian Experience* (New York: Paulist Press, 1974).

In the first chapter of his book, Lee sets the stage for his process-oriented understanding of Christ, church, and sacraments. He takes note, for example, of the phenomenon of secularization within Western civil society and by extension within the church as well.[3] This was in part an indirect consequence of modern interest in natural science with its focus on empirical data but also the result of contemporary biblical criticism, subjecting the text of Sacred Scripture to rational analysis.[4] Still another factor in the phenomenon of secularization was the popularity of post–World War II existentialist philosophy with its emphasis on human freedom, the ability to become whatever one wished to be, free of restraints in terms of God's will or natural law.[5] To meet the challenge thus presented to the church by contemporary Western culture, concludes Lee, theologians must look for new symbols and new modes of thought, chief among them the process philosophy of Whitehead, to better express the ancient faith.[6]

Lee then sketches different versions of a new process-oriented Christology. The first, largely based on the thought of John Cobb, employs the Whiteheadian concepts of divine initial aim and the subjective aim of the concrescing actual entity to explain how Jesus was different from other human beings, even from the Hebrew prophets, in his personal life and message. Since this is basically the same as what I outlined in chapter nine, I will not mention any further details. The second process-oriented Christology is the work of Pierre Teilhard de Chardin, the Jesuit priest/paleontologist who in his philosophical/theological writings reconciled his scientific belief in cosmic evolution with the traditional teaching of the Roman Catholic Church about the divinity of Jesus and, above all, about the Second Coming of Jesus at the end of the world.[7] Teilhard was not a process philosopher, still less a student of Whitehead, but his writings reflect a basically process-oriented approach. Yet one must never forget the differences between Teilhard and Whitehead—above all, in what they say about Jesus as the Christ. Here Teilhard, thinking more as a theologian than as a philosopher, feels free to use the New Testament to support his belief in the Cosmic Christ as the Omega or endpoint of cosmic evolution. Whitehead, thinking more as a philosopher than as a theologian, foresees

3. Ibid., 10–12.
4. Ibid., 20–22.
5. Ibid., 22–28.
6. Ibid., 43–52.
7. See Pierre Teilhard de Chardin, *The Phenomenon of Man*, trans. Bernard Wall (New York: Harper & Row, 1963).

158   *Part Three: Christian Doctrinal Questions*

no end to the cosmic process as a whole but only an end to what he calls a series of cosmic epochs.

The worldview or metaphysical vision of the two men, however, is strikingly similar. Both agree "that the world is both radically particulate and essentially related."[8] Likewise, both seek to overcome the dichotomy between matter and spirit in scientific research and in religious practice. Teilhard, accordingly, postulates that all energy is basically psychic but is at work in nature under two different forms: tangential energy, which links together entities of the same size and complexity so as to constitute the "Without" or external appearance of things, and radial energy, which draws an individual entity toward ever-greater complexity and centricity and thus constitutes its "Within."[9] Whitehead comes basically to the same conclusion with his notion of an actual entity. An actual entity is internally an immaterial subject of experience and externally a material reality, aspects of whose objective pattern of self-constitution can be physically "prehended" by subsequent actual entities.[10] Moreover, the action of God in the world is quite similar for both men. "Teilhard also sees the action of God as a kind of lure. With something of a dipolar sense of God, he speaks of evolution as 'stirred by the Prime Mover ahead.'"[11] For both Whitehead and Teilhard, accordingly, God and creativity, the urge for novelty and progress, are closely linked. Likewise, as Lee points out, the basic understanding of the supernatural in Teilhard's work is also strongly process oriented. That is, the supernatural represents not what is above and beyond the cosmic process but what lies ahead in the cosmic process.[12]

The chapters on church and sacraments are the heart of Lee's analysis of the becoming of the church. In one chapter, for example, he applies Whitehead's generic description of society as an assemblage of actual entities, momentary self-constituting subjects of experience, linked by a "common element of form" to the church as a "structured society" or society composed of subsocieties of actual entities.[13] Lee begins his explanation by asking himself what the common element of form should be for Christians as members of the church and concludes that it is "the

---

8. Ibid., 127–28. See also Whitehead, *Process and Reality*, 18.

9. Teilhard, *The Phenomenon of Man*, 129–30.

10. Alfred North Whitehead, *Process and Reality: An Essay in Cosmology*, corrected edition, ed. David Ray Griffin and Donald W. Sherburne (New York: Free Press, 1978), 27–28.

11. Lee, *The Becoming of the Church*, 135. The reference is to Teilhard's *The Phenomenon of Man*, 271.

12. Lee, *The Becoming of the Church*, 139–40, 151.

13. Ibid., 174–81.

Jesus-event."[14] Later in the book he specifies more carefully what he means by the Jesus-event. But initially he is simply interested in the dynamics of church life, how the Jesus-event is what links Christians to one another as members of the church. In line with Whitehead's understanding of society, he concludes: "The defining characteristic of a society emerges in an individual through certain conditions imposed upon him by his prehensions of those who are already members of the society."[15] That is, new members come to understand and assimilate the Jesus-event by noting how it functions within the lives of fellow members of the church. But once the new member of the church appropriates the Jesus-event as constitutive of his or her own existence, he or she thereby reinforces that common element of form for other church members, both new and old. Both the individual and the group in this way "emBody" the Jesus-event, make it present to one another within the church and to the world at large.[16]

Lee then invokes John Cobb's understanding of the structure of Christian existence: "Christian existence is 'defined as spiritual existence that expresses itself in love.'"[17] Not just any kind of love is herewith implied, of course, but only "a love which frees us from needing to be loved by anyone else [so as] to know that we are loved, and frees us to love more largely than ever."[18] Jesus, in other words, both in his life and in his message, affirmed the unconditional love of God for human beings, based upon his own experience of God the Father's unconditional love for him even under the trying circumstances of his passion and death on the cross.[19] Thus understood, the Jesus-event transforms Christian life to make it the Sacrament of God's love: in the first place for church members in their dealings with one another, but in the second place for church members in their dealings with the non-Christian world where skepticism and even hostility toward this kind of self-giving love is all too often experienced.[20]

Finally, Lee applied to the church as a Whiteheadian structured society the same goals of group survival and intensity of experience for component parts or members that would be applicable to biological organisms within the cosmic process.[21] That is, in order to survive within an ever-changing

---

14. Ibid., 175–76.
15. Ibid., 176. See also Whitehead, *Process and Reality*, 34.
16. Lee, *The Becoming* of the *Church*, 177–79.
17. Ibid., 183. See also John B. Cobb Jr., *The Structure of Christian Existence* (Philadelphia, PA: Westminster Press, 1967), 125.
18. Lee, *The Becoming of the Church*, 183.
19. Ibid., 185.
20. Ibid., 186.
21. See Whitehead, *Process and Reality*, 100–101.

environment, physical organisms have to be relatively unspecialized—have a simple structure or pattern of existence. But to further evolve and grow, an organism must have component parts or members that in their dynamic interrelation exhibit intensity of experience. Lee comments, "As a society the Church meets the same challenge, to maintain itself in existence and to occasion an intense life in the membership."[22] For this to happen, what Whitehead calls "conceptual reversion" must be at work in the minds and hearts of church members. Conceptual reversion is the way in which contemporary members of a society reshape the common element of form or "eternal object" governing their dynamic interrelation in order to adjust to novelty within the environment.[23] In Lee's view, this is the privileged role of the sacraments in church life. "Each Sacramental experience is a possible initiative 'to receive the novel elements of the environment into explicit feelings with such subjective forms [emotional overtones] as conciliate them with the complex experiences proper to members of the structured society.'"[24] In this way, sacraments for Lee are events, not things. Likewise, the church is not so much a stable institution with a fixed body of beliefs and practices, but an ongoing process constituted again and again by sacramental events.[25]

To explain sacraments as sacred events, not sacred things, Lee first appeals to the exodus event in the Hebrew Bible and the annual celebration of Passover every spring by the Jewish people. "The event was not completed once and for all. It continues to take hold of each man's life."[26] The annual Passover celebration has made that identity-shaping event present to Jews ever after. In similar fashion, the sacraments allow Christians to prehend the Jesus-event as something both past and present, something "felt *there* and made to be *here*."[27] That "something" in Whiteheadian terms is an "eternal object," an existential possibility that was objectively realized or made actual in the life and message of Jesus and that subjectively affects a Christian in his or her self-constitution here and now. Lee cautions, however, that the institutional church must carefully supervise this symbolic reference of the Jesus-event from the past into the present.[28] Otherwise, there is clear danger that some Christians will unconsciously

---

22. Lee, *The Becoming of the Church*, 197.
23. Whitehead, *Process and Reality*, 102.
24. Lee, *The Becoming of the Church*, 200. See also Whitehead, *Process and Reality*, 102.
25. Lee, *The Becoming of the Church*, 205–7.
26. Ibid., 211.
27. Ibid., 212.
28. Ibid., 213, 220.

misread the significance of the original Jesus-event and make it present to themselves here and now in the wrong way with negative consequences for themselves and others.

Yet in each sacramental experience there is also need for intensity dynamics as well as for survival dynamics. Intensity dynamics are concerned with the here-and-now impact of the Jesus-event on the church member.[29] A sacramental ritual, however well performed, only makes a difference to an individual Christian if it brings home the meaning and value of the Jesus-event to him or her in the present moment. In this sense, the traditional preoccupation of the Roman Catholic Church with the matter and the form of the sacrament is largely misplaced. The sacrament is not a thing composed of matter and form but an event designed to make a difference in the lives of Christians sharing in the event.[30] Appealing to Whitehead's description of how an actual entity is influenced by the society to which it belongs and by the external environment, Lee states, "Many prehensions go into a concrescence [the internal growth of an actual entity]. Each prehension has a subjective form, that is, its own emotional tone. Each emotion is a response to some quality or another (forms of definiteness, eternal objects)."[31] Sacramental rituals, in other words, have different effects on different people. The ritual, after all, only partially objectifies the Jesus-event for any given Christian. Both as a historical and as a contemporary reality, the Jesus-event exceeds the capacity of the ritual to express and of the participant to comprehend. But if the ritual is properly performed, some specific dimensions (forms of definiteness) of the Jesus-event will be conveyed with sufficient emotional power so as to catch the attention of the participants and bring home to them the importance of the Jesus-event for their own self-constitution or self-identity here and now. This seems to correspond to a time-honored distinction in classical sacramental theology between *ex opere operato* and *ex opere operantis* (between the ritual itself and the one who performs or goes through the ritual). The objective long-term efficacy of the sacrament is heavily dependent upon the readiness of the individual Christian to act upon the grace of the sacrament.

Lee analyzes three sacraments (baptism, confirmation, and eucharist) from a process- or event-oriented perspective. In each case, he makes clear the indispensable role of the Christian community in conveying and sustaining the importance of the Jesus-event for the individual believer. With respect to baptism, for example, the person to be baptized or person newly

---

29. Ibid.
30. Ibid., 223.
31. Ibid., 226–27.

baptized needs to prehend the Jesus-event at work in other members of the Christian community (in the first place, family members) in order to begin seeing himself or herself as likewise molded by the Jesus-event. "His fundamental understandings are given him by the community in which he finds life. His own sense of identity is shaped by those around him."[32] As Lee comments, this is akin to the first stage in the self-constitution of an actual entity, concrescence, where it is heavily conditioned by other actual entities in its environment. With respect to confirmation, the person baptized is "confirmed" by the bishop or other church authorities as "one to whom others ought to be able to look to see a workable 'pattern of assemblage' according to which a Christian life has been put together."[33] This, in turn, corresponds to the second stage in the self-constitution of an actual entity, that of superject, where the actual entity's pattern of self-constitution is available for prehension by subsequent actual entities for their self-constitution.

But it is primarily in Lee's analysis of the Eucharist as a sacred event that the community plays an all-important role. Lee first calls attention to the fact that in Paul's account of the first Eucharist in 1 Corinthians 11:23-26, the consecration of the bread and the wine into the Body and Blood of the Lord do not take place at the same time during the paschal meal: "The cup is taken only after the meal. . . . The breaking of the bread is earlier in the meal."[34] In Luke's account, there are two cups shared, one during the meal and one at the end; only the latter cup is offered as Christ's Blood to be shed in testimony to the new covenant.[35] The practice of the early church, however, followed the account of the institution of the Eucharist in Matthew and Mark, where the consecrated bread and wine are shared in quick succession during the meal. Lee prefers the institution narrative in Paul and Luke since it allows him to claim that the consecration of the bread and the wine are different symbolic actions (events) with related but still different meanings.[36] "It seems clear to me that the Sacramental symbols, as reported in Paul's letter (and for the most part, also in Luke's account) are not *things* of bread and wine, though they are part, but the *action* of breaking bread and partaking, and of blessing the cup of a new covenant in Jesus' blood and partaking in it."[37]

32. Ibid., 236.
33. Ibid., 242.
34. Ibid., 246–47.
35. Ibid., 247.
36. Ibid.
37. Ibid., 248.

So the classical doctrine of transubstantiation, whereby bread and wine are transformed into the Body and Blood of the Lord, has a misplaced emphasis. The emphasis should be not on a miraculous change of one substance into another but on two powerful community-building actions or events: breaking consecrated bread and sharing consecrated wine, which remind those who participate that they now belong to the new covenant, the new relationship of humanity with God established by Jesus through his life and message, but above all through his passion and death on the cross. The particular context for Jesus' institution of a new covenant at the Last Supper, of course, was the Passover, the yearly ritual reenactment of the exodus from Egypt, which in Jewish belief forever establishes their covenant with God through Moses. Here too bread is broken and shared, and blood is shed, as key symbols of the enduring covenant of God with his chosen people. Lee then comments, "In a way, the content of Jesus' New Covenant is not greatly different. The locus of it is in the spirit of love and not in prescriptions of a written law. . . . Jesus does, however, un-nationalize the Covenant. It is meant for all men. It is universalized. Hebrew history erupts into world history in the Jesus-event."[38] In both cases, of course, the covenant is a community-building event; it builds the community of God with human beings and the community of human beings with one another in their concelebration of a ritual meal.

Finally, in a chapter entitled "Some Pastoral Implications," Lee reviews the ways in which a process-oriented approach to church and sacraments should make a major difference in the lives of ordinary Christians. He begins by noting that becoming a Christian is not a one-time event but a lifelong process of assimilating features of the Jesus-event into one's own ongoing self-constitution or self-identity: "The name Christian belongs to an identity which is larger than individual actions, which is as large as patterns of life, a pattern which inundates our process, communicating its identity to the whole assemblage of life's events."[39] This is not to say that all Christians will assimilate the Jesus-event in the same way. Because different Christians will incorporate different dimensions of the Jesus-event into their lives, there is still another reason for life in Christian community. Through their prehending one another's different personal assimilations of the Jesus-event, individual Christians will over time gain a certain balance in the way they conduct themselves before the world as Christians.[40]

---

38. Ibid., 252.
39. Ibid., 257.
40. Ibid., 258–60.

Unquestionably, Lee's use of Whiteheadian metaphysics to interpret what it means to be a Christian in the modern world has much to recommend it. His event- or process-oriented approach to Christian life makes membership in the church much more dynamic for the individual believer. One feels much more connected with the Jesus-event both as a historical event and as an ongoing reality shaping one's own life and the life of others within the Christian community. One sees much better how Christians need one another to understand and interpret the deeper significance of the Jesus-event for their lives together in community. Christian community is thus not only a matter of going to church on Sundays and putting an envelope in the collection basket to pay one's share of necessary expenses. It is much more a matter of Christians becoming together the Body of Christ in this place and at this time in cosmic history. At the same time, Lee's approach with its focus on the "how" as opposed to the "what" of the Jesus-event seems to prescind from the concrete details of life within each one of the different Christian denominations. The process of internalizing the Jesus-event and becoming church together with other like-minded Christians would seem in Lee's account to apply indiscriminately to Roman Catholics, Baptists, Methodists, Presbyterians, etc. In one respect, this may be an advantage. But in another respect, it seems to ignore the top-down causation of a distinctive ecclesial environment upon the bottom-up causation of the individual and his or her faith community in interpreting and acting out the Jesus-event here and now. But if there are real differences in the understanding and administration of church and sacraments among the various Christian denominations, can one remedy that "defect" and still retain a basically Whiteheadian metaphysics as one's governing conceptuality for rethinking church and sacraments in contemporary life?

How, for example, would my own revised understanding of Whiteheadian societies as structured fields of activity for their constituent actual entities be of use to retain a process-oriented approach to church and sacraments? First of all, I suggest that we think of Christianity as a complex Whiteheadian structured society with the different Christian denominations as its constituent subsocieties.[41] Each of these subsocieties is a structured field of activity for its church members (Roman Catholics, Episcopalians, Methodists, etc.). Hence, each of these worshipping communities will cherish its own traditional understanding of church and sacraments. Yet each of them also contributes to the common element of

---

41. See Joseph A. Bracken, *Christianity and Process Thought: Spirituality for a Changing World* (West Conshohocken, PA: Templeton Foundation Press, 2006), 56.

form or "defining characteristic" of the larger structured society that is Christianity as a whole. Hence, Christianity cannot be defined simply in terms of what it means to be Roman Catholic, Baptist, or Presbyterian. All of them help to present Christianity as a complex institutional reality to the outside world.

Yet Christianity is only one of the major world religions. As I will explain in more detail in chapter twelve, interreligious dialogue would be enormously enhanced if the leaders of all the world religions would concede that their religious community is a subsociety within the all-encompassing structured society that in biblical terms may be called the kingdom of God or in more neutral terms the community of world religions. For the moment, however, what I am striving for in this chapter is a way from a Whiteheadian perspective to justify what I earlier called top-down causality as well as bottom-up causality within each of the various Christian denominations. Each of these subfields of ecclesial activity within the broader field of activity proper to Christianity as a whole has an abiding internal structure that is consistent with its institutional history but that makes it different from the other Christian denominations with their own structured fields of activity for their members.

So the overall unity of Christianity or the Christian religion is a differentiated rather than a uniform unity. From a philosophical perspective, this is still another verification of my proposed paradigm for the relation between the One and the Many. The One is not a higher-order institutional entity that brings about coherence and unity among the different Christian denominations in virtue of its own unchanging reality. Rather, the One is emergent out of the ongoing interplay of the Many, the different Christian denominations, in dealing with issues of common concern. As a result, Christianity as a world religion will evolve over time, take on a new configuration as its parts or members shift in their relations to one another and to the circumstances of an ever-changing external world. In particular, while the Christian sacraments should everywhere be understood to function as events, not as things, in line with Lee's proposal, their specific reality as sacred events will be somewhat different within each Christian denomination, each subfield of ecclesial activity. For example, both the number of sacraments (seven for Roman Catholics, only two or three for many others) and their accompanying ceremonies differ from one denomination to another so that the Jesus-event is not experienced in quite the same way by Christians belonging to different denominations. This is perfectly consistent with a field-oriented approach to Whiteheadian structured societies whereby each subsociety has its own ontological identity and yet can coparticipate with other subsocieties in perpetuating the meaning and value of the broader structured society.[42] Applied to the Jesus-event, this means that the various Christian denominations need one another to give full expression to their ongoing corporate reality as the Body of Christ. They share with one another the ideal of unity in diversity as the best way to express to the world that transcendent reality.

---

42. Ibid., 59–61.

12

Inclusivity and Exclusivity in a Religious Context

In the preceding chapters of this book, I have argued that a new paradigm for the relation between the One and the Many is slowly gaining ground in Western culture under the impact of an evolutionary / systems-oriented approach to reality, namely, an understanding of the One as emergent out of the ongoing dynamic interplay of the Many with one another.[1] The One, accordingly, is no longer a transcendent entity but a structured field of activity or common space for the Many in their ever-changing relations to one another. Such a philosophical paradigm can be derived, as Colin Gunton notes, from the notion of *perichoresis* for the understanding of the doctrine of the Trinity among the early Greek fathers of the church.[2] It can also be derived from a rethinking of the category of "society" in the philosophy of Alfred North Whitehead.[3] That is, a society should not be understood simply as an aggregate of actual entities, momentary self-constituting subjects of experience, with a "common element of form" or analogous self-constitution,[4] but rather as a structured

---

1. The historical background in the history of Western philosophy for this rethinking of the classical relation between the One and the Many has, of course, been provided in my earlier publication *Subjectivity, Objectivity, and Intersubjectivity: A New Paradigm for Religion and Science* (West Conshohocken, PA: Templeton Foundation Press, 2009).

2. Colin E. Gunton, *The One, the Three and the Many: God, Creation and the Culture of Modernity* (Cambridge: Cambridge University Press, 1993), 163–78.

3. Joseph A. Bracken, *Subjectivity, Objectivity and Intersubjectivity: A New Paradigm for Religion and Science* (West Conshohocken, PA: Templeton Foundation Press, 2009), 124–37.

4. Alfred North Whitehead, *Process and Reality: An Essay in Cosmology*, corrected edition, ed. David Ray Griffin and Donald W. Sherburne (New York: Free Press, 1978), 34.

field of activity or common space for those same actual entities in dynamic interrelation.[5] In either case, the point is that the One is no longer a transcendent entity but a structured environment or context for the ongoing relations of the Many to one another.

In the present chapter, I apply this new paradigm for the dynamic interrelation of the One and the Many to three progressively broader religious contexts: the relation within Roman Catholicism between the papacy, understood as the universal church, and all the particular dioceses / regional churches scattered around the world; the relations between all the various churches or ecclesiastical denominations within Christianity as a world religion; and finally, the relations of the various world religions to one another in terms of contemporary interreligious dialogue.

The One and the Many within Roman Catholicism

For some background to this first task, namely, the relation between the pope and the other Roman Catholic bishops around the world, I turn to Bernard Prusak's *The Church Unfinished*.[6] In this book he argues that in the Dogmatic Constitution on the Church (*Lumen Gentium*), the bishops at the Second Vatican Council, without denying the primacy of the pope and the church in Rome, nevertheless tried to recover the earlier understanding of the universal church as a "communion of churches."[7] To legitimate that claim, he reviews the institutional growth of the church from the apostolic period onward and traces the gradual growth in the influence and authority of the bishop of Rome as the successor to Peter the apostle. This growth in influence, of course, should not be seen simply as a "power grab" on the part of the medieval popes but as an honest effort to protect the independence of the church from the meddling of the nobility into the selection of bishops and therewith control over church properties in a given area. With the Protestant Reformation in the sixteenth century, of course, the prestige and, above all, the legislative power of the pope was seriously called into question.[8] But the subsequent Counter-Reformation of the Catholic Church, while it instituted some much-needed reform in the training of the clergy and administration of local dioceses, still strongly maintained the primacy of the pope and the Vatican bureaucracy in the

---

5. Bracken, *Subjectivity, Objectivity, and Intersubjectivity*, 129–30.
6. Bernard P. Prusak, *The Church Unfinished: Ecclesiology through the Centuries* (New York: Paulist Press, 2004).
7. Ibid., 120–30.
8. Ibid., 235–47.

life of the church. In that sense, the decree *Pastor Aeternus* approved by the bishops at the First Vatican Council, which provided for the personal infallibility of the pope under special conditions in matters of faith and morals, was a natural consequence of the continued focus on central authority within the Roman Catholic Church.[9]

With this historical background, we have the setting for the documents of Vatican II, especially the one on the internal organization and structure of the church, *Lumen Gentium*. The Vatican bureaucracy and conservatively oriented bishops wanted a reaffirmation of the personal infallibility of the pope in matters of faith and morals. But liberally oriented bishops wanted more emphasis on their own authority as bishops, especially when gathered together as an ecumenical council. The result was a compromise position, articulated in chapter three of *Lumen Gentium* as follows:

> The Roman Pontiff, as the successor of Peter, is the perpetual and visible source and foundation of the unity of the bishops and of the multitude of the faithful. The individual bishop, however, is the visible principle and foundation of unity in his particular church, fashioned after the model of the universal Church. In and from such individual churches there comes into being the one and only Catholic Church. For this reason, each individual bishop represents his own church, but all of them together in union with the Pope represent the entire Church joined in the bond of peace, love, and unity.[10]

Upon closer analysis, this citation from *Lumen Gentium* contains the two different models for the relation between the One and the Many cited above. The Vatican bureaucracy and conservatively oriented bishops were reaffirming the necessity for the unity of the church in a higher-order entity, the pope as the Vicar of Christ. The liberally oriented bishops were urging another model in which the unity of the church was cocreated by the dynamic interrelations of the bishops with one another throughout the world. Yet the liberals also wanted to preserve in some fashion the historical privileged place for the bishop of Rome within the hierarchy.

Not surprisingly, in the decades since the close of Vatican II, there has been a major growth of interest within the Roman Catholic Church in "communion ecclesiology" as a way to reaffirm the primacy of the pope as defined by Vatican I and at the same time to set forth a stronger understanding of the church as a communion of bishops around the world

---

9. Ibid., 256–57.
10. *Lumen Gentium*, no. 23, in *Documents of Vatican II*, ed. Walter M. Abbott, SJ (New York: Guild Press, 1966), 44.

170 *Part Three: Christian Doctrinal Questions*

acting in conjunction with the pope and the Vatican for the direction of the whole church. In this chapter I cannot explore that line of thought as it was set forth by an extraordinary synod of bishops in 1985 in a document under the title The Church, in the Word of God, Celebrates the Mysteries of Christ for the Salvation of the World, likewise in the document issued by the Congregation for the Doctrine of the Faith in 1992 with the title Some Aspects of the Church Understood as Communion, and thirdly in the celebrated debate in separate issues of *America* magazine in 2001 between then Cardinal Ratzinger (now Pope Benedict XVI) and Cardinal Kasper over the issue of the proper relation between the universal church and all the local churches around the world.[11] But what seems to have emerged as a common theme in all three contexts is a tension between rival understandings, first, of what is meant by the term "communion" and, secondly, of what is meant by the term "universal church." With respect to the notion of communion, Ratzinger and other more traditionally oriented theologians prefer to think of the church as mirroring the communion of the divine persons with one another in their intratrinitarian life. As a result, their notion of the universal church tends to be somewhat ahistorical, even mystical, a Platonic ideal more than a historical reality.[12] For Kasper and other more liberally oriented theologians, the notion of communion is verified, at least in part, in the existing relationship of the pope and bishops to one another within the college of bishops, the relation of bishops to their brother priests as pastors of local parishes, and the relationship of priest-pastors with their parishioners on a daily basis as together constituting a worshipping community. As a result, the understanding of the term "universal church" for Kasper and others is closer to what the council fathers of Vatican II in the citation from *Lumen Gentium*

---

11. See here Kilian McDonnell, OSB, "The Ratzinger/Kasper Debate: The Universal Church and Local Churches," *Theological Studies* 63 (2002): 227–50. McDonnell offers a detailed analysis of each of these documents in terms of the proper relationship between the universal church and the local churches. Likewise, see Edward P. Hahnenberg, "The Mystical Body of Christ and Communion Ecclesiology: Historical Parallels," *Irish Theological Quarterly* 70 (2005), 3–30. Hahnenberg provides a historical overview of the various models for the Church that appeared in the twentieth century, thus from before Vatican II, during Vatican II, and after Vatican II. His conclusion is that none of them fully encompass the divine-human reality of the Church and thus were sooner or later supplemented by still other models. The model of the Church as communion, accordingly, will in due time be supplemented by other models still not fully articulated.

12. Cf. Hahnenberg, "The Mystical Body of Christ and Communion Ecclesiology, 20–21.

identified as the "entire Church," the communion of churches around the world with one another and with the church in Rome.[13]

If we look at the matter in terms of the problem of the One and the Many, the group headed up by then Cardinal Ratzinger and his followers tend to think of the One, namely, the unity of the church, in a strictly entitative sense, that is, as focused in the Roman Church as the indispensable principle of unity and cohesion for all the other churches scattered around the world, which, taken together, are the Many. The other group led by Cardinal Kasper and his followers see the One or the ongoing unity of the Catholic Church not as an institutional entity but as a communion or structured field of activity for the dynamic interaction of pope and bishops with one another, of bishops with their pastors, of pastors with their parishioners. For Kasper and his colleagues, accordingly, the unity of the Catholic Church is that of a Whiteheadian structured society or society of subsocieties, each with its own integrity and mode of operation, but all of them contributing to the overall reality of the universal church as a structured society. From my own perspective, the church as a Whiteheadian structured society is an integrated set of ecclesiastical fields of activity ordered to one another both vertically and horizontally. But if this is indeed the case, then all along in their polite exchanges with one another, there were deep philosophical differences between Ratzinger, Kasper, and their respective followers about the meaning of the term "communion" and about the nature of the universal church. Is there any chance of a compromise position to be worked out between the two groups whereby each one could claim that its basic premise is in the end vindicated?

In line with the basic thrust of this book, I claim that a compromise position is possible, at least in principle, if one properly understands what Whitehead had in mind with the notion of a structured society, namely, a society made up of subsocieties in dynamic interrelation. That is, if each of the bishops in the Roman Catholic Church, including the pope, can be thought of as presiding over a given subsociety within the universal church as a Whiteheadian structured society, and if these dynamically interrelated subsocieties function together properly, then it is altogether legitimate for

---

13. Cf. McDonnell, OSB, "The Ratzinger/Kasper Debate," 246–50. See also Gerard Mannion, *Ecclesiology and Postmodernity: Questions for the Church in Our Time* (Collegeville, MN: Liturgical Press, 2007), esp. 44–74. In these pages Mannion sums up the philosophical and theological presuppositions of the position of Cardinal Joseph Ratzinger (now Pope Benedict XVI) and more conservatively oriented bishops and theologians. In the rest of the book he outlines the basic features of the notion of communion ecclesiology espoused by Karl Rahner, Gregory Baum, Roger Haight, and others (including himself).

one of those subsocieties to play a leading role within the structured society as a whole. Such a very large and complex structured society presumably would have need of one of those subsocieties to function as the principle of order and unity for all of them if they are to stay together as a unified corporate reality. This one subsociety would be akin to the notion of "soul" in classical metaphysics, except that such a higher-order subsociety normally takes shape only over time in virtue of the interplay of all the subsocieties as they reach a certain level of organization and complexity. Hence, this higher-order society, unlike the "soul" in classical metaphysics, is not different in kind from the other subsocieties but only different in terms of its ordering function within the entire group of subsocieties. Furthermore, its very existence depends upon its ongoing interdependence with the other subsocieties. Translated into more commonsense language, the pope within this theoretical scheme would equivalently be the CEO or executive officer of the college of bishops around the world. In this capacity, the pope would serve as the principle of coherence and order for the universal church but would not have the unilateral legislative power of the office of the papacy as it has developed from medieval times until the present day.

Naturally, such a rethinking of the role of the papacy in the Roman Catholic Church would probably not be agreeable to Pope Benedict XVI, the Vatican bureaucracy, and more conservatively oriented bishops because, at the very least, it would reduce the church to a purely human organization, devoid of all the mystery attached to a reality instituted by God from all eternity.[14] But, as Prusak points out in *The Church Unfinished*, it would seem to correspond rather closely to the actual practice of the early church.

During the second and third centuries, provincial and regional synods of bishops were convened to resolve the controversies concerning the date of Easter and the problems posed by Montanism. By the fourth and fifth centuries, the *koinonia* or communion of the universal church had come to mean the unity of churches that were not sectarian in their doctrine or discipline. The universal communion of such churches was mediated by the communion of their bishops, who now also assembled for general or ecumenical councils, convoked by the Christian emperors, to respond to serious threats to the unity of the faith and of the newly Christian empire.[15]

At the same time, of course, the church of Rome was seen as the center of unity for this communion of bishops. For example, Prusak cites

---

14. *Lumen Gentium*, nos. 1–8, in *Documents of Vatican II*, ed. Walter M. Abbott, SJ, 14–24.

15. Prusak, *The Church Unfinished*, 130.

Inclusivity and Exclusivity in a Religious Context 173

to that effect Irenaeus, bishop of Lyons at the end of the second century: "The bishop of Rome taught in a way that reflected the genuine apostolic faith of all the churches, because this church gathered together or was in dialog with persons from many different parts of the world. Thus, by listening to Rome, one heard the faith of the entire Church."[16] Likewise, for Augustine in the fifth century, "the bishop of Rome represented the unity of the Church as a whole, as Peter had. Peter's successors taught what the Roman Church had always held in harmony with the faith of all the other churches, whether Latin or Greek, in the only truly catholic or universal Church."[17] So there is ample historical precedent for seeing the pope in the early church as presiding over the college of bishops but not as governing it through strictly unilateral action. In any event, I offer this modified understanding of the role of the papacy within the college of bishops as an option for the pope and bishops to discuss and evaluate at some future date (perhaps at a Third Vatican Council?).

The One and the Many within Ecumenical Discussion

Can this same revised understanding of a Whiteheadian "structured society" likewise serve as a model for the relations of all the Christian churches to one another? That is, if Christianity as a world religion is understood as a Whiteheadian structured society or complex structured field of activity emergent out of the ongoing ecumenical exchange between different Christian denominations as its constituent subsocieties, then perhaps there is a way for each of these denominations to be itself and yet to be defined in its self-identity through its relations to the other Christian denominations. After all, Whitehead's notion of a society as constituted by the way in which its constituent actual entities or self-constituting subjects of experience are *internally* rather than *externally* related to one another makes clear the necessary interdependence of the various Christian denominations with one another as they together bring to self-expression the corporate reality of Christianity from moment to moment. In support of this proposal, one need only make reference here to the classical Thomistic understanding of the Trinity in which the three divine persons are regarded as "subsistent relations," defined in their individual self-identity by their internal relations to one another as Father, Son, and Holy Spirit.[18] That

---

16. Ibid., 134.
17. Ibid., 139.
18. Cf. Thomas Aquinas, *Summa Theologica*, trans. Fathers of the English Dominican Province (New York: Benzinger Bros., 1947), I, q. 29, a. 4, resp.

is, they represent three totally different ways of being God, and yet only in and through these very differences are they the one God.

In any case, such a theoretical model for the relations of the various Christian denominations to one another requires that each denomination see itself in terms of its relations to all the other denominations.[19] One does not know, for example, what it means to be Roman Catholic except insofar as one acknowledges both the similarities and the differences between the doctrines and liturgical practices of the Catholic Church and those of all the other Christian denominations. The same, of course, would be true of all the Protestant denominations in their relations to one another, to the Roman Catholic Church, and to the various branches of Eastern Orthodox Christianity. The fullness of what it means to be Christian is only achieved when one takes into account the way in which the Gospel teachings have been instantiated in all the different Christian churches.

Does such an ecclesiastical reconfiguration, however, require a higher-order principle of unity or executive director such as happens with the office of the papacy within Roman Catholicism? In principle such a communion of different Christian denominations could exist as a pure democracy with total equality in its mode of operation. But there are distinct advantages to having a chief spokesperson for any large group in dealing with still other groups. The key point, of course, is that the spokesperson, like the pope or possibly the executive director of the World Council of Churches, would see his or her spiritual authority as ultimately derivative from the consent of the various Christian denominations rather than as permanently resident in an individual with the unilateral power to impose a given doctrine or liturgical practice on all the members. If such an individual would make a public statement for the entire group, it would presumably get the attention of the international news media in much the same way that the pope does at the present time. So there are good pragmatic reasons for preserving something like the spiritual authority of the pope or some other ecclesiastical spokesperson to present a united Christian front to the representatives of the other world religions and to the public at large through the news media.

The One and the Many within Interreligious Circles

Finally, I believe that this model of a Whiteheadian society as an enduring structured field of activity for its constituent actual entities could

---

19. Cf. Joseph A. Bracken, *Christianity and Process Thought: Spirituality for a Changing World* (Philadelphia, PA: Templeton Foundation Press, 2006), 59–61.

also be applied to the issue of the proper relation of the world religions to one another.[20] Since some of the world religions like Buddhism and Taoism are nontheistic and others like classical Hinduism are seemingly polytheistic, adjustments have to be made in terms of how to label the common space created by the ongoing exchange between representatives of all the religious traditions. I would suggest the neutral term "Ultimate Reality," which can be applied to whatever the representatives of a given world religion hold to be sacred. In that case, all the world religions can share with one another their understanding of what is for them "ultimate reality" and thus by implication their "ultimate concern"[21] or raison d'être. For theistic religions like Judaism, Christianity, and Islam, such a conversation should have the effect of broadening their understanding of God as a truly multidimensional reality. That is, the God of Jews, Christians, and Muslims can be conceived as in different contexts unipersonal, tripersonal, or transpersonal (namely, the Godhead, which as a nonpersonal reality has some affinity with the notion of Brahman, Emptiness, and the Tao in east Asian religions[22]). Nontheistic religions like Buddhism and Taoism or a religion with a supposed pantheon of gods like classical Hinduism can come to appreciate the need for a stronger ethical dimension for their otherwise more individualistically oriented east Asian religious traditions. The Semitic religions (Judaism, Christianity, and Islam), in other words, have from the beginning insisted that God is intimately involved in human history. Hence, as part of the covenant relationship with God, believers are expected to serve the needs of others, not only fellow Jews, Christians, and Muslims, but all human beings around the world as children of God, made in the image and likeness of God and thus as worthy of inclusion in the kingdom of God, both here and hereafter. Furthermore, as a result of such ongoing interreligious exchange, the world religions should be better able to combat materialism in all its various forms and to promote a common sense of transcendent values necessary for the future well-being of the human race and indeed of the earth itself as a fragile ecological system.

To sum up, my argument in this chapter has been that all the above-mentioned changes in attitude among religiously oriented people could be significantly helped by recognizing the real differences between the two rival paradigms for understanding the proper relationship between of the One and the Many, which I have set forth at some length in this

---

20. Ibid., 61–64.
21. Cf. Paul Tillich, *The Dynamics of Faith* (New York: Harper & Row, 1957), 1–4.
22. See Joseph A. Bracken, *The Divine Matrix: Creativity as Link between East and West* (Maryknoll, NY: Orbis Books, 1995), 73–127.

book. After all, the more aware one is of these key differences in outlook and mode of operation, the more one is equipped to decide when the one paradigm is clearly better than the other in a given practical situation. In other words, awareness of the differences between the two models of reality does not mean that one is ultimately forced to make a one-time, either-or choice. There are times and places when the classical paradigm with its hegemony of the One over the Many fits the bill perfectly. An example would be the institution of the papacy within the Roman Catholic Church. There are other times and places where the alternate paradigm with its understanding of the One as emergent out of the interchange of the Many with one another within a common space or structured field of activity resolves a lot of turf issues: relations of the different world religions to one another, relations of the various Christian denominations to one another, and within Roman Catholicism the relation between the pope and his brother bishops around the world. I myself generally favor the more process-oriented approach to the relation between the One and the Many in this and other instances because it seems more open to adaptation and further development than the classical Platonic model for the One and the Many. But given the complexity of modern life, there is no one-size-fits-all solution to the never-ending tension between unity and plurality, identity and difference, in the way we human beings have to deal with one another.

Conclusion

Granville Henry notes in an early chapter of his book *Christiainity and the Images of Science* that Pythagoras and his school were not acquainted with the possibility of fractions as multiple ways to subdivide the number one without losing the sense of "one" as the first prime number. For them, "one" was the only number that was simple and undifferentiated; all other numbers like "two" were composed of parts—for example, two "ones."[1] According to Henry, this was the origin of the common belief among the philosophers of antiquity that perfection is to be found in what is simple, without differentiation into parts. What has parts, after all, can fall apart and would have to be put together again by an outside agency. For that matter, whatever has parts necessarily owes its current existence to an outside agency. As Aristotle phrased it, whatever is moved is moved by another.[2] From here it was an easy inference for Thomas Aquinas and other medieval thinkers to think of God as in the first place simple, without parts.

Aquinas, to be sure, in his description of the three divine persons as together one God, one corporate reality, had in his hands by implication a new understanding of unity or the One as unity in diversity of parts or members. But insofar as he implicitly separated his understanding of the doctrine of the Trinity from his overall philosophical scheme for the God-world relationship in the *Summa Theologica*,[3] for Aquinas the notion of

---

1. Granville C. Henry, *Christianity and the Images of Science* (Macon, GA: Smyth & Helwys, 1984), 40–41.
2. Aristotle, *Physics*, bk. VII, chap. 1 (241b24), in *The Basic Works of Aristotle*, ed. Richard McKeon (New York: Random House, 1941), 340.
3. Thomas Aquinas, *Summa Theologica*, trans. Fathers of the English Dominican Province (New York: Benzinger, 1948), I, qq. 27–43.

the One as undifferentiated or without parts remained unchallenged. In similar fashion, through much of the early modern period the philosophical notion of the One as undifferentiated unity was not challenged. As I have pointed out elsewhere, reference to the One shifted from the Creator God of biblical revelation to the individual Self as ordering principle of its multiple experiences.[4] But the underlying presupposition that the One is a higher-order individual entity that transcends the Many as their ordering principle remained unchanged. The individual Self was assumed to transcend everything that it experienced. Similarly, through the early modern period political life in the West was dominated by the idea that the One (e.g., the king) must be above the Many (the ordinary citizens) to properly govern the State. Only in the bold experiment with more democratic forms of government in the American and French Revolutions of the late eighteenth century did the idea of democracy in political life begin to take hold in the minds of people around the world. Yet even then some measure of elitism remained. In the United States, for example, for many years after the War of Independence only white, male property owners could vote and run for office. Likewise, looking around the world in the twenty-first century, one quickly sees that a democratic way of life in civil society remains more an ideal to be striven for than an actual achievement.

Yet at least in academic circles, the philosophical notion of the One as a differentiated as opposed to an undifferentiated unity has been steadily gaining ground. In the natural sciences, for example, a delicate balance between top-down and bottom-up causation, the reciprocal influence of the One on the Many and the Many on the One, seems to be more and more accepted by scientists. The strict reductionists are certainly correct in affirming the need for bottom-up causation at all levels of existence and activity within nature, but top-down causality, at least in the form of "information," seems to be also necessary for the understanding of how order evolves out of apparent chaos. Thus, as these natural scientists see it, the whole is in some way more than simply the sum of its parts. A newly emergent form of order and intelligibility within an organism seems to set constraints on the continued activity of the organism's component parts or members even as the component parts or members provide the infrastructure for the original emergence and continued survival of the higher-order form of existence and activity for the organism as a whole. Nor is this to be understood as an implicit reaffirmation of the distinction between

---

4. Joseph A. Bracken, *Subjectivity, Objectivity, and Intersubjectivity: A New Paradigm for Religion and Science* (West Conshohocken, PA: Templeton Foundation Press, 2009), 126.

matter and form in classical metaphysics. Matter in classical metaphysics is purely passive; form alone is responsible for the internal organization of the entity. With this proposed understanding of the interplay between bottom-up and top-down causation, both matter and form are actively at work within the entity but in different ways.

Then, shifting to the social sciences, one notes how systems thinking has become a generally accepted methodology for scientific research. One sees the world as hierarchically ordered in terms of systems and subsystems.[5] But as I indicated in chapter five, one way to understand what is meant by "system" is to see it as still another exemplification of this new paradigm for the relationship between the One and the Many. The constituent parts or members of the system at every moment contribute to the functional unity of the system as an organized whole. Just as in the natural sciences, therefore, both bottom-up and top-down causation are involved in the organization and ongoing operation of social systems, albeit in different ways. To belong to a system, potential parts or members have to adjust to the constraints built into the configuration of the system as a whole. Yet the system itself would quickly break down without the ongoing interrelated activity of its component parts or members.

The notion of system as a differentiated unity of dynamically interrelated parts or members, therefore, largely seems to have replaced the classical notion of substance and accident for the understanding of what is going on in nature within both the natural and social sciences. What I have added over and above this generic understanding of system is to claim that the best way to explain philosophically what is meant by the term is to envision a system as a Whiteheadian "society," more specifically as a structured field of activity or law-like environment for the ongoing dynamic interrelation of its component parts (actual entities or momentary subjects of experience). Through its governing structure or "common element of form," the society / structured field of activity exercises top-down formal (or informational) causality upon its constituent actual entities, even as these same components by their interaction both with one another and with the external environment provide bottom-up causation to preserve the ongoing structure of the society and to provide for a gradual change in that structure in response to external environmental factors. Thus, as Niklas Luhmann pointed out in chapter seven, a system is a basically self-sufficient reality, constantly adjusting to its external environment in virtue of its own internal resources.

---

5. Ervin Laszlo, *Introduction to Systems Philosophy: Toward a New Paradigm of Contemporary Thought* (London: Gordon and Breach, 1972), 47–53.

Finally, within the field of Christian doctrinal theology, this understanding of a Whiteheadian society as a structured field of activity for its constituent actual entities at any given moment throws unexpected light on traditional Christian beliefs that were always hard to explain philosophically, still more so from a scientific perspective. Belief that Jesus of Nazareth is a divine person with both a divine and a full human nature, for example, has been a puzzle for Christian theologians since the official proclamation of the doctrine at the Council of Chalcedon in 451.[6] But if the divine persons are one God in virtue of coconstituting a shared field of activity as a result of their ongoing relations to one another, then it makes perfect sense to claim that Jesus as a divine person exists in both the unlimited divine field of activity that he shares with the Father and the Spirit and in the finite field of activity proper to his human nature. The fields are interrelated but still distinct from one another. The human field of activity is necessarily incorporated into the infinite reality of the divine field of activity but still exists in its own right as a semiautonomous finite field of activity. Thus, the laws proper to the functioning of systems in the natural and social sciences can be generally applied to the understanding of the life and work of Jesus in his human nature or finite field of activity, but they do not apply to what Jesus as a divine person does or can do in terms of his divine nature or divine field of activity (especially after his death and resurrection).

In similar fashion, one can possibly reconcile scientific understanding of the laws of nature with Christian belief in the resurrection of the body after death. That is, if one thinks of human nature in each of us as an ongoing field of activity that has been progressively structured by all the events taking place within us and around us in the course of our lives, then one can first stipulate that this extensive but still quite finite field of activity has over a lifetime been progressively incorporated into the prevailing structure of the divine field of activity, or what I call in biblical terms the kingdom of God. That is, just as Jesus in his human nature added to the fullness of the divine life by his human life, death, and resurrection, so each of us through the prevailing pattern of our lives on earth have contributed to the fullness of the divine life, the kingdom of God as an all-encompassing structured field of activity that serves as the shared "space" within which the divine persons can communicate with all their creatures—above all, with their human creatures. Then at the moment of death, each of us through a

---

6. *Enchiridion Symbolorum, Definitionum et Declarationum de rebus fidei et morum*, ed. Henricus Denziger and Adolfus Schönmetzer, SJ (Freiburg: Herder, 1973), nos. 301–2.

special grace or favor from the divine persons will be privileged to "see" for the first time in its fullness what we have contributed to the divine life, the kingdom of God, within our own lifetime. We will see ourselves as still the same bodily individuals that we have always been but in a new light.

With respect to church and sacraments, one can also use the notion of a Whiteheadian society as a structured field of activity for its constituent actual entities to describe how Jesus sought to bring about the longed-for kingdom of God by consciously setting up a new subfield of activity, a small community of his disciples who would carry on his mission in life after his untimely death and departure from this life. That is, within the all-encompassing field of activity proper to the kingdom of God as the generic "space" for the never-ending interaction between God and creatures, Jesus established a new subfield of activity with quite specific goals and values. Through the Eucharist, the ritual commemoration of the Last Supper, his disciples were to celebrate both the death of Jesus as atonement for their sins and the totally unexpected gift of his resurrection as testimony to the possibility of a new way of life in Christ through the power of the Holy Spirit. New converts to this grace-filled way of life were, moreover, to be inducted into the community through baptism, the ritual expression of death to an older sinful way of life and of resurrection to the new grace-filled way of life initiated by Jesus and afterwards faithfully practiced by his followers. Yet as a finite field of activity within the infinite field of activity proper to the kingdom of God, the church should also be capable of analysis by the various social sciences. Obviously, the social sciences do not have the last word in evaluating the origin and ongoing history of the church, but they should have something valuable to say about its current oversight and management by human beings. It is a human as well as a divine institution.

For similar reasons, the notion of a Whiteheadian society as a field of activity structured by the dynamic interrelationship of its constituent actual entities from moment to moment could be quite useful for analyzing and effectively dealing with issues of authority within the Roman Catholic Church—for example, the relation between the Pope and the bishops, between the bishops and the priests together with the people in the parishes that they are ordained to serve. This same heuristic structure might likewise be useful in settling differences among the various Christian denominations in their common task of preaching the Gospel and in setting up a more favorable atmosphere for dialogue among the major world religions in their conjoint attempts to deal with common concerns within the modern world. With respect to the issue of authority within the Roman Catholic Church, of course, it is helpful to remember that, while Whiteheadian societies are radically egalitarian in the way that their constituent actual entities in equal measure influence one another's self-identity, Whitehead also allows for structured societies, that is, societies composed of subsocieties, that have a regnant subsociety to provide order and direction to the other subsocieties. Hence, the leadership role of the papacy can be justified even within a communion model of the bishops in their interactions with one another. In terms of the ecumenical dialogue between the various Christian denominations, one can draw on the standard Whiteheadian doctrine that the constituent actual entities of a society have internal, and not purely external, relations to one another. Each Christian denomination has something distinctive to give to the ongoing ecumenical discussion and something equally distinctive to learn from that discussion for its own self-identity and way of procedure. Finally, all the major world religions might find it easier to unite in setting up an all-embracing religiously oriented field of activity immanent within and yet transcendent of the material conditions of life. In this way, they can together address the challenge of a purely secular and often grossly materialistic way of life that threatens the common good both of the human race and of the entire planet in terms of its ecological well-being.

To sum up, then, this book claims that a significant shift in worldview or perspective on life is basically at stake in what might seem to be at first glance a rather trivial change in one's abstract understanding of the proper philosophical relation between the One and the Many. For the reconception of the One not as a higher-order entity but as a common structured field of activity for the dynamic interrelationship of the Many with one another is by implication the philosophical basis for a new socially oriented worldview in which individuals find their deeper identity in contributing to a socially organized reality much bigger than themselves simply as individuals. Such an ontology would, at the very least, be an effective remedy for the exaggerated emphasis on the individual and individual rights that has developed over the centuries within Western civilization and that is now taken for granted by many people as simply the way things are. Yet we contemporary human beings live de facto in a deeply interconnected world in which, through the internet and other telecommunications media, we share the same unsettling experiences of a rapidly changing world as people who are immediately affected by the events in question. So we are sorely in need of a theoretical justification for us to live together more amiably and thus with less fear and anxiety precisely as members of a world community with multiple overlapping subsocieties or dynamically interrelated organizational systems (political, economic, religiously oriented, and purely social institutions). Finally, since the earth itself is a delicately balanced ecological network of plants, animals, and human beings in which each living species plays a distinctive role in maintaining the well-being of the entire system, there is even more need for a new paradigm for the proper relation between the One and the Many that emphasizes the significance and value of a shared life with others even at the risk of some personal loss as a result of working unselfishly for the common good.

Selected Bibliography

Books

Aquinas, Thomas. *On the Power of God*. Translated by the English Dominican Fathers. Westminster, MD: Newman, 1952.

———. *Summa Theologiae*. Madrid, Spain: Biblioteca de Autores Cristianos, 1955.

———. *Summa Theologica*. Translated by the Fathers of the English Dominican Province. New York: Benziger, 1948.

Aristotle. *The Basic Works of Aristotle: Metaphysics*. Edited by Richard McKeon. New York: Random House, 1941.

Barbour, Ian G. *Religion and Science: Historical and Contemporary Issues*. San Francisco, CA: HarperCollins, 1997.

Behe, Michael J. *Darwin's Black Box*. New York: Free Press, 1996.

Bracken, Joseph A. *Christianity and Process Thought: Spirituality for a Changing World*. Philadelphia, PA: Templeton Foundation Press, 2006.

———. *The Divine Matrix: Creativity as the Link between East and West*. Maryknoll, NY: Orbis Books, 1995.

———. *God: Three Who Are One*. Collegeville, MN: Liturgical Press, 2008.

———. *The One in the Many: A Contemporary Reconstruction of the God-World Relation*. Grand Rapids, MI: Eerdmans, 2001.

———. *Subjectivity, Objectivity and Intersubjectivity: A New Paradigm for Religion and Science*. West Conshohocken, PA: Templeton Foundation Press, 2009.

———. *The Triune Symbol: Persons, Process and Community*. Lanham, MD: University Press of America, 1985.

Clayton, Philip. *Mind and Emergence: From Quantum to Consciousness*. Oxford, UK: Oxford University Press, 2004.

Cobb, John B., Jr. *The Structure of Christian Experience*. Philadelphia, PA: Westminster Press, 1967.

Collins, James. *A History of Modern European Philosophy*. Milwaukee, WI: Bruce Publishing, 1954.

Darwin, Charles. *The Descent of Man and Selection in Relation to Sex.* New York: Appleton, 1871.
Dawkins, Richard. *The Selfish Gene.* Oxford, UK: Oxford University Press, 1976.
Dembski, William A. *The Design Inference: Eliminating Chance through Small Possibilities.* Cambridge, UK: Cambridge University Press, 2001.
Depew, David J., and Weber, Bruce H. *Darwinism Evolving: Systems Dynamics and the Genealogy of Natural Selection.* Cambridge, MA: Massachusetts Institute of Technology Press, 1995.
Derrida, Jacques. *Margins of Philosophy.* Translated by Alan Bass. Chicago, IL: University of Chicago Press, 1982.
———. *Of Grammatology.* Translated by Gayatri Chakravorty Spivak. Baltimore, MD: Johns Hopkins University Press, 1976.
Descartes, René. *The Philosophical Works of Descartes.* 2 vols. Translated by Elizabeth S. Haldane and G. R. T. Ross. Cambridge, UK: Cambridge University Press. 1978.
Documents of Vatican II. Edited by Walter M. Abbott, SJ. New York: Guild Press, 1966.
Edwards, Denis. *How God Acts: Creation, Redemption and Special Divine Action.* Minneapolis, MN: Fortress Press, 2010.
Enchiridion Symbolorum, Definitionum et Declarationum de rebus fidei et morum. Edited by Henricus Denzinger and Adolfus Schönmetzer, SJ. Freiburg, Germany: Herder, 1973.
Felt, James W. *Coming to Be: Toward a Thomistic-Whiteheadian Metaphysics of Becoming.* Albany, NY: State University of New York Press, 2004.
Fiddes, Paul. *The Creative Suffering of God.* Oxford, UK: Clarendon Press, 1988.
Forman, Robert K. C., ed. *Religions of the World.* Third Edition. New York: St. Martin's Press, 1993.
Gaillardetz, Richard. *Ecclesiology for a Global Church: A People Called and Sent.* Maryknoll, NY: Orbis Books, 2008.
Gerhart, Mary, and Allan Melvin Russell. *New Maps for Old: Explorations of Science and Religion.* New York: Continuum, 2001.
Gilson, Etienne. *The Unity of Philosophical Experience.* Westminster, MD: Four Courts Press, 1982.
Gunton, Colin E. *The One, the Three and the Many: God, Creation and the Culture of Modernity.* Cambridge, UK: Cambridge University Press, 1993.
Haught, John F. *God after Darwin: A Theology of Evolution.* Boulder, CO: Westview Press, 2000.
Heidegger, Martin. *Being and Time.* Translated by John Macquarrie and Edward Robinson. New York: Harper & Row, 1962.
Henry, Granville C. *Christianity and the Images of Science.* Macon, GA: Smyth & Helwys, 1998.
Hobbes, Thomas. *Leviathan, or the Matter, Forme and Power of a Commonwealth, Ecclesiastical and Civil.* Edited by Michael Oakeshott. Oxford, UK: Blackwell, 1960.

Hume, David. *A Treatise of Human Nature*. Edited by I. A. Selby-Biggs. Oxford: UK: Clarendon Press, 1967.
Johnson, Elizabeth A. *The Mystery of God in Feminist Theological Discourse*. New York: Crossroad, 1992.
Jones, W. T. *A History of Western Philosophy*. Second edition. New York: Harcourt, Brace and World, 1969.
Julian of Norwich. *Showings*. Edited by Edmund Colledge and James Walsh. New York: Paulist Press, 1978
Kant, Immanuel. *Immanuel Kant's Critique of Pure Reason*. Translated by Norman Kemp Smith. New York: St. Martin's Press. 1964.
Kauffman, Stuart. *At Home in the Universe: The Search for the Laws of Self-Organization and Complexity*. New York: Oxford University Press, 1995.
———. *Investigations*. New York: Oxford University Press, 2000.
———. *Reinventing the Sacred: A New View of Science, Reason, and Religion*. New York: Basic Books, 2008.
Kierkegaard, Søren. *Concluding Unscientific Postscript to Philosophical Fragments*. 2 vols. Translated by Howard V. Hong and Edna H. Hong. Princeton, NJ: Princeton University Press, 1992.
———. *Fear and Trembling*. Translated by Howard V. Hong and Edna H. Hong. Princeton, NJ: Princeton University Press, 1983.
Korsmeyer, Jerry D. *Evolution and Eden: Balancing Original Sin and Contemporary Science*. New York: Paulist Press, 1998.
Kushner, Harold. *When Bad Things Happen to Good People*. New York: Schocken Books, 2004.
Laszlo, Ervin. *The Connectivity Hypothesis: Foundations of an Integral Science of Quantum, Cosmos, Life, and Consciousness*. Albany, NY: State University of New York Press, 2003.
———. *Introduction to Systems Philosophy: Toward a New Paradigm of Contemporary Thought*. London: Gordon and Breach, 1972.
———. *The Systems View of the World: The Natural Philosophy of the New Development in the Sciences*. New York: Braziller, 1972.
Leclerc, Ivor. *The Nature of Physical Existence*. New York: Humanities Press, 1972.
———. *The Philosophy of Nature*. Washington, DC: Catholic University of America Press, 1986.
Lee, Bernard. *The Becoming of the Church: A Process Theology of the Structure of Christian Experience*. New York: Paulist Press, 1974.
Levinas, Emmanuel. *Totality and Infinity: An Essay on Exteriority*. Translated by Alphonso Lingis. Pittsburgh, PA: Duquesne University Press, 1969.
Locke, John. *An Essay Concerning Human Understanding*. Edited by Peter Niddich. Oxford, UK: Clarendon Press, 1975.
———. *An Essay Concerning the True Original, Extent and End of Civil Government*. In *The English Philosophers from Bacon to Mill*, edited by Edwin Burtt. New York: Modern Library, 1939.

Loy, David. *Nonduality: A Study in Comparative Philosophy.* New Haven, CT: Yale University Press, 1988.
Luhmann, Niklas. *Social Systems.* Translated by John Bednarz, Jr. with Dirk Baecker. Stanford, CA: Stanford University Press, 1996.
Lyotard, Jean-Francois. *The Postmodern Condition: A Report on Knowledge.* Translated by Geoff Benningon and Brian Massumi. Minneapolis, MN: University of Minnesota Press, 1984.
Mannion, Gerard. *Ecclesiology and Postmodernity: Questions for the Church in Our Time.* Collegeville, MN: Liturgical Press, 2007.
McFague, Sallie. *Models of God: Theology for an Ecological Nuclear Age.* Philadelphia, PA: Fortress Press, 1987.
Mitchell, Donald W. *Buddhism: Introducing the Buddhist Experience.* New York: Oxford University Press, 2002.
Moeller, Hans-Georg. *Luhmann Explained: From Souls to Systems.* Chicago, IL: Open Court, 2006.
Neville, Robert Cummings. *Eternity and Time's Flow.* Albany, NY: State University of New York Press, 1993.
Nicholas of Cusa. *Trialogus de Posset.* In Jasper Hopkins, *Introduction to the Philosophy of Nicholas of Cusa.* Minneapolis, MN: University of Minnesota Press, 1978.
Niebuhr, Reinhold. *The Nature and Destiny of Man: A Christian Interpretation.* New York: Harcourt, Brace and World, 1941.
Pagels, Heinz. *The Cosmic Code: Quantum Physics as the Language of Nature.* New York: Bantam Books, 1984.
Peirce, Charles Sanders. *Collected Papers of Charles Sanders Peirce.* Vols. 5 and 6. Edited by Charles Hartshorne and Paul Weiss. Cambridge, MA: Harvard University Press, 1934 and 1935.
Plato. *The Republic.* Translated by Francis MacDonald. New York: Oxford University Press, 1962.
Polkinghorne, John. *The God of Hope and the End of the World.* New Haven, CT: Yale University Press, 2003.
Prusak, Richard. *The Church Unfinished: Ecclesiology through the Centuries.* New York: Paulist Press, 2004.
Rahner, Karl. *Foundations of Christian Faith: An Introduction to the Idea of Christianity.* Translated by William V. Dych. New York: Crossroad, 1978.
———. *Theological Investigations.* Vol. 7. Translated by Cornelius Ernst. London: Darton, Longman & Todd, 1971.
Richard, Louis. *The Mystery of the Redemption.* Translated by Joseph Horn. Baltimore, MD: Helicon Press, 1965.
Rousseau, Jean Jacques. *The Social Contract and Discourses.* London: Everyman's Library, 1923.
Royce, Josiah. *The Problem of Christianity.* Chicago, IL: University of Chicago Press, 1968.

Ruse, Michael. *Can a Darwinian be a Christian? The Relationship between Science and Religion*. Cambridge, UK: Cambridge University Press, 2001.
Schelling, F. W. J. *Schellings Werke*. Vol. 11. Edited by Manfred Schröter. Munich: Germany C. H. Beck. 1968.
Smith, Huston. *Forgotten Truth: The Primordial Tradition*. New York: Harper & Row, 1976.
———. *The World's Religions* [originally *The Religions of Man*]. New York: HarperCollins, 1991.
Sober, Elliott, and David Sloan Wilson. *Unto Others: The Evolution and Psychology of Unselfish Behavior*. Cambridge, MA: Harvard University Press, 1998.
Southgate, Christopher. *The Groaning of Creation: God, Evolution, and the Problem of Evil*. Louisville, KY: Westminster/John Knox, 2008.
Spinoza, Benedict. *Spinoza's Ethics and De Intellectus Emendatione*. Translated by Andrew Boyle. London: J. M. Dent, 1959.
Suchocki, Marjorie Hewitt. *The End of Evil: A Process Eschatology*. Albany, NY: State University of New York Press, 1985.
Teilhard de Chardin, Pierre. *The Phenomenon of Man*. Translated by Bernard Wall New York: Harper and Row, 1965.
Tillich, Paul. *The Dynamics of Faith*. New York: Harper & Row, 1957.
Tipler, Frank. *The Physics of Immortality: Modern Cosmology, God, and the Resurrection of the Dead*. New York: Doubleday, 1994.
Ward, Keith. *The Big Questions in Science and Religion*. West Conshohocken, PA: Templeton Press, 2008.
———. *Pascal's Fire: Scientific Faith and Religious Understanding*. Oxford, UK: One World, 2006.
Weber, Michel. *Whitehead's Pancreativism: Jamesian Applications*. Frankfurt am Main, Germany: Ontos Verlag, 2011.
Whitehead, Alfred North. *Adventures of Ideas*. New York: The Free Press, 1967.
———. *Process and Reality: An Essay in Cosmology*. Corrected edition. Edited by David Ray Griffin and Donald W. Sherburne. New York: Free Press, 1978.
———. *Science and the Modern World*. Second edition. New York: Macmillan, 1967.
Wiehl, Reiner. *Subjektivität und System*. Frankfurt am Main, Germany: Suhrkamp, 2001.
Wilson, David Sloan. *Darwin's Cathedral: Evolution, Religion, and the Nature of Society*. Chicago, IL: University of Chicago Press, 2003.
Wink, Walter. *Engaging the Powers: Discernment and Resistance in a World of Domination*. Minneapolis, MN: Fortress Press, 1992.
———. *Naming the Powers: The Language of Power in the New Testament*. Philadelphia, PA: Fortress Press, 1984.
———. *Unmasking the Powers: The Invisible Forces that Determine Human Existence*. Philadelphia, PA: Fortress Press, 1986.

Zycinski, Josef. *God and Evolution: Fundamental Questions of Christian Evolutionism*. Translated by Kenneth W. Kemp and Zusanna Maslanka. Washington, DC: The Catholic University of America Press, 2006.

Articles and Book Chapters

Bracken, Joseph A. "Response to Elizabeth Johnson's 'Does God Play Dice?'" *Theological Studies* 57 (1996): 720–30.

Brooke, John Hedley. "Natural Theology." In *Science and Religion: A Historical Introduction*, edited by Gary Ferngren. Baltimore, MD: Johns Hopkins University Press, 2002.

Caponi, Francis J. "Pale Analogies and Dead Metaphors." *Horizons* 37 (2010): 37–42.

Griffin, David Ray. "Of Minds and Molecules." In *The Reenchantment of Science: Postmodern Proposals*, edited by David Ray Griffin. Albany, NY: State University of New York Press, 1988.

Hahnenberg, Edward. "The Mystical Body of Christ and Communion Ecclesiology." *Irish Theological Quarterly* 70 (2005): 3–30.

Hartshorne, Charles. "The Compound Individual." In *Philosophical Essays for Alfred North Whitehead*, edited by F. S. C. Northrup. New York: Russell & Russell, 1936.

Johnson, Elizabeth A., CSJ. "Does God Play Dice? Divine Providence and Chance." *Theological Studies* 57 (1996): 3–18.

McDonnell, Kilian, OSB. "The Ratzinger/Kasper Debate: The Universal Church and Local Churches." *Theological Studies* 63 (2002): 227–50.

McHenry, Leemon B. "Maxwell's Field and Whitehead's Events: The Adventure of a Revolutionary Idea." In *Subjectivity, Process and Rationality*, edited by Michel Weber and Pierfrancisco Basile. Frankfurt am Main, Germany: Ontos Verlag, 2007.

Quine, Willard. "Whither Physical Objects?" In *Essays in Memory of Imre Lakatos*, edited by R. S. Cohen, P. K. Feyerabend, and M. W. Wartofsky. Boston Studies in the Philosophy of Science 39. Dordrecht, Netherlands: D. Reidel, 1976.

Rapp, Frederich. "Whitehead's Concept of Creativity and Modern Science." In *Whitehead's Metaphysics of Creativity*, edited by Reiner Wiehl and Frederick Rapp. Albany, NY: State University of New York Press, 1990.

Van Till, Howard. "Dialogues: Is Naturalistic Christianity the Way to Go? A Response to David Griffin." *Theology and Science* 2 (2004): 175–81.

Weber, Michel. "Introduction: Process Metaphysics in Context." In *After Whitehead: Rescher on Process Metaphysics*, edited by Michel Weber. Heusenstamm bei Frankfurt, Germany: Ontos Verlag, 2004.

Wiehl, Reiner. "Whitehead's Cosmology between Ontology and Anthropology." In *Whitehead's Metaphysics of Creativity*, edited by Reiner Wiehl and Frederick Rapp. Albany, NY: State University of New York Press, 1990.

Index

Abelard, Peter, 132, 138
actual entity/actual occasion, xvi, 7,
    11, 18, 21–29, 36, 38–40, 42,
    43–44, 53–56, 71–74, 76–77, 80–
    83, 89, 92, 94–100, 103, 106–9,
    115–17, 119–23, 128–30, 134,
    140–43, 146–48, 151, 158, 161,
    164, 167–68, 173, 179–82
  internal vs. external relations, 56,
    182–83
  living/non-living, 82, 86–87
  physical/mental pole, 82, 147–48
  self-constituting decision, 18,
    21–22, 27–29, 42, 43, 55–56,
    71, 81–83, 95–96, 107, 109,
    129–30, 134, 138, 145–46
actuality vs. potentiality, 16, 18–19,
    24, 83, 96, 98, 100
aesthetics vs. ethics, 19–20, 29, 66
agency
  of actual occasions, 39–40, 73–74,
    77, 84, 88–89, 99, 121–22
  of societies, 39–40, 72–74, 84,
    86–89, 99, 121–22
aim, divine initial, 128–30, 157
  subjective, 128, 157
altruism, 76, 81–83
Aquinas, Thomas, xii, 16–17, 24, 32,
    47, 50–52, 55, 61–62, 73–74, 88,
    132, 155, 173, 177

Anselm of Canterbury, 132
appearance vs. reality, 139, 146,
    152–53, 166
Aristotle, xii, 16, 23, 51–52, 61–62,
    67, 73–74, 86, 88, 99, 141, 177
Atomism, 84–85, 95, 97–98, 151
atoms
  material, 7–8, 10, 39, 52, 71, 77,
    80, 85, 117, 140
  spiritual, 7–8, 10, 21, 39, 53, 71,
    77, 80, 85, 117, 140
Augustine, 31
authority
  in public life, 111–24, 178
  in religious circles, 167–76, 181
autocatalytic mechanism, 40
autopoiesis, 102, 105, 107

Barbour, Ian, 52
Bernard of Clairvaux, 132
Big Bang, 11, 24–25, 48, 144
body vs. soul/mind, 65, 86–89, 127,
    139–53
Bracken, Joseph A.
  Christianity and Process Thought,
    xiii, 10, 12, 22
  Subjectivity, Objectivity and Intersubjectivity, xiii, 22, 25–26, 61,
    63–66, 68
  The Divine Matrix, 24

191

The One in the Many, xii, 9

causal efficacy, 8
causality
   bottom-up and top-down, 26, 44–45, 72–73, 99, 107, 120, 156, 164, 178–79
   formal/informational, 142
   primary and secondary, 43, 47–58, 62
cause-and-effect relationship, xii, 4, 32, 53, 62–64
chance/indeterminability, 4, 6, 13, 21, 34, 42, 44, 48, 80, 104
chaos-and-order, 3–4, 6–7, 9, 31–45, 93, 101–2, 178
Christ. *See* Jesus the Christ
Christianity vs. Judaism, 133
Church, Christian, 134, 137, 155–66, 168, 173–74, 181–82
   Roman Catholic, 137, 155–66, 168–76, 181
   Universal Church, 170, 171–73
Clayton, Philip, xii
Cobb, John, 143, 156–57, 159
Collins, James, 114
communion ecclesiology, 169–70, 182
community/civil society
   divine, 112, 129, 144
   human, 114–15, 121–22, 132–33, 135–38, 149, 161–65, 182–83
consciousness/intentionality, 79, 81–83, 104, 129–30, 139, 142, 146, 151
contingency, ix–x, 48, 54, 70
cosmic vacuum/plenum, 118–19
councils of the Church, 172–73
   Chalcedon, Council of, 131, 180
   Ephesus, Council of, 127
   First Vatican Council, 169
   Nicaea, Council of, 127
   Second Vatican Council, 137, 156, 168–71
Creatio ex nihilo, 11, 144
creativity, 15–29, 54, 80–82, 84, 93, 101, 143, 158

Darwin, Charles, 17, 19, 32–33, 47, 75
Dawkins, Richard, 75, 80
democratic process, 111–23, 178
Derrida, Jacques, 67–68, 91, 94
Descartes, René, 63, 74, 113
Determinism, ix–x, 4–6, 32, 35, 48, 54
Dualism vs. Monism, 26, 45, 56, 61, 65–66, 158

Edwards, Denis, 47–58
Einstein, Albert, ix–x, 118
emergence, xii, 43, 50
eschatology, 50–52, 150–52, 157
Eternal Object, 95, 169
eternity vs. time, 152
ethics vs. aesthetics, 19–20, 29, 35–36, 66
events
   energy-events, 4, 8, 118
   events vs. things, 160–65
   Jesus-event, 159–66 (*see also* Jesus the Christ)
   mental events, 6
   psycho-physical events, 7–8
evil vs. good, 21, 35–36, 50, 54–55, 131–32, 134, 136–38, 144–45, 147
evolution, goals of, 6
   trial-and-error process, 13, 48–49, 51, 80
exclusivity/inclusivity. *See* inclusivity/exclusivity

field of activity, structured, xii, xiii n.10, 9–12, 23–24, 42, 43–45, 56–57, 71, 73–74, 76, 85, 88, 92, 99–100, 103, 106–9, 117–21, 128, 133–35, 137–38, 141–46, 148–51, 164–65, 167–68, 171, 173, 179–82
Foucault, Michel, 91
freedom, human, 13, 22, 130–31, 157
fundamentalism, 4

Gilson, Etienne, 54, 103

God
　as Creator, 19–20, 27, 34, 52, 55, 61, 63, 113, 133, 144, 158, 178
　as multi-dimensional, 175
　as simple (without parts), 177
　consequent nature of God, 29, 42, 43, 94–96, 108, 142, 144, 151
　Divine Attractor, 31–36
　Divine Communitarian Life, 56, 143–44, 146, 152–53, 170, 180–81 (see also kingdom of God)
　divine matrix/divine field of activity as ground of creation, 24, 29, 45, 144, 146
　divine omnipotence, 3–4
　divine providence, 13, 28–29, 32–36, 42–44, 48–58, 148
　divine transcendence/immanence in creation, 55, 58, 144
　kingdom of God, 10–12, 20, 29, 48, 108, 128, 133–34, 137–38, 145, 149, 153, 165, 175, 180–81
　mercy vs. justice of God, 132, 138
　primordial nature of God, 28–29
　subsistent being/supreme actuality, 15, 24, 62, 101
God-world relationship, x, xii, 3–4, 10–13, 15, 18, 23, 28–29, 31–36, 45, 48–49, 54, 62, 71, 91, 93, 113, 127–38, 142–45, 177, 182
good vs. evil, 21, 54–55, 131–32, 134, 136–38, 144–45
　as collective realities, 134–38
Gunton, Colin, 112–14, 167

habits/habit-making, 5–6, 21, 64, 135
Hartshorne, Charles, 87–88, 143
heaven, 146
Hegel, G. W. F., 65–66, 70, 92, 96
Heidegger, Martin, 67, 101
　Dasein, 101
hell, 147

Henry, Granville, x, 177
Heraclitus, 112–13
Hobbes, Thomas, 114
Hume, David, 53, 64

inclusivity/exclusivity, 167–76
Information Theory, 118–20, 178–79
initial aims, divine, 11–12, 22, 27–28, 36, 40, 42–43, 54, 96, 146
intelligent design, 31–45
interreligious dialog, 155–56, 165, 174–75
intersubjectivity, 48, 52–53, 55, 58, 66, 101, 104–6, 123

Jesus the Christ, x, 6, 12, 20, 33, 35, 48–50, 127–38, 146, 149–50, 157, 181
　death of, 131, 133, 149–50, 181
　incarnation, doctrine of, 127–38, 180
　Jesus-event, 159–66
　Mystical Body of, 140, 145, 149, 153, 163, 166
　redemption through, 20, 52, 55–56, 131–38
　resurrection of, 50, 129, 133, 149–50 (see also resurrection)
　risen body of, 148–53
　sinlessness of, 129, 137–38
Johnson, Elizabeth, xv n. 15, xvi n. 16
Judaism, 160, 163
Julian of Norwich, 13

Kant, Immanuel, 64–65, 74, 104
Kasper, Walter Cardinal, 170–71
Kauffman, Stuart, xii, 15–16, 18, 31–32, 36–42
Kierkegaard, Søren, 6
Kushner, Harold, 3, 20

Last Judgment, x, 150–52
laws of nature, 3–4, 9, 32, 41, 48
　statistical, 32
　universal, 32, 34, 48, 50, 64

Laszlo, Ervin, 71–74, 97, 99–100, 118–21
  natural vs. artificial systems, 71, 99–100
Leclerc, Ivor, 97–99
  compound individuals, 98, 100
Lee, Bernard, 155–66
Leibniz, Gottfried, 97
  Monads, 98
Levinas, Emmanuel, xi, 67, 91
life after death, 56, 139–53 (*see also* resurrection)
Locke, John, 63–64, 114, 122–23
logocentrism, 94
love, self-sacrificing, 6, 159, 183
Luhmann, Niklas, xiv, 92, 102–8, 123, 179
  general systems theory, 102
  structural coupling, 102
  system differentiation, 107–8
Lyotard, Jean-Francois, xi, 67

Mary, Mother of God, 129–30
McFague, Sallie, 143
mechanistic world view, 85
meta-narrative, xi, 74
mind/nature, law of, 5, 9
miracles, 50
Mitchell, Donald W., 23
models, use of, 52, 57–58, 176
Moeller, Hans Georg, 105, 108
Monism vs. Dualism, 25–26, 45, 56, 158, 175
mystery, divine, 48, 58

naturalism
  methodological, 32–33
  ontological, 32–33, 45, 76
natural selection, 17, 19, 36–37, 75, 80–81, 89, 117, 148, 153
natural theology, x
nature
  self-organizing, 9, 31–45, 75–89
Newtonian mechanics, ix, 85

nexus vs. society, 86–87, 94
Nicholas of Cusa, 34
Niebuhr, Reinhold, 136
non-dualism, 26

One and the Many, xi–xiii, 88, 95–96, 100, 111–24, 133, 155–56, 165, 167–69, 171, 175–76, 177–83
ontological principle, Whitehead's, 93
order vs. chaos/novelty, 3–4, 6–7, 9, 31–45, 93, 101, 102, 178

Pagels, Heinz, ix
panentheism vs. pantheism, 34, 45, 127, 142–44
panpsychism, 26
papal primacy/infallibility, 168–74, 176, 181
Parmenides, 112–13
parts vs. the whole, 141, 145–46
Paul, the Apostle, 134, 138, 145, 162
Peirce, Charles Sanders, 4–7, 10, 21, 26
persons
  divine, xvi, 10, 12, 24, 35, 44–45, 48, 57, 108, 112, 127–38, 143, 146–47, 149–50, 159, 170, 173, 175, 177, 180–81
  human, 8, 127–38, 144–45, 149, 159
Pope Benedict XVI (Cardinal Ratzinger), 170–72
Plato, xii, 61, 67, 74, 88, 113, 122, 136, 170, 176
Plotinus, 16
potentiality vs. actuality, 16, 18–19, 83, 96, 100
power, coercive vs. persuasive, 48
prehension, 79, 82, 94, 99, 115, 142, 145, 161
presentational immediacy, 8
probability theory, x, 33–35
proposition
  logical, 82
  Whiteheadian, 82–83, 93

Protestant Reformation and Catholic Counter-Reformation, 168
Prusak, Bernard, 168, 172–73
purgatory, 147
Pythagoras, 177

quantum mechanics, ix, 54, 118, 153
Quine, Willlard, 118

Rahner, Karl, 47, 50–51
Rapp, Friedrich, 92, 101
reality vs. appearance, 139–40, 146, 152–53, 166
reason vs. revelation, 58, 153 (*see also* religion and science debate)
redemption/salvation, theories of, 131–38, 139–53, 180–81
reductionism, 178 (*see also* Monism vs. Dualism)
regularity vs. uniformity, 6, 9
religion and science debate, xi, xiii, 76 (*see also* reason vs. revelation)
resurrection of the body (*see also* life after death)
    Jesus, 50, 129, 133, 180–81 (*see also* Jesus the Christ)
    human beings, 139–53, 180
    creation as a whole, 142–44, 148
Rousseau, Jacques, 114–15, 122–23

sacraments of the church, 155–66, 181
    baptism, 161–62, 181
    confirmation, 162
    Eucharist, 162–63, 181
    transubstantiation, 163
salvation history, 132
Schelling, F. W. J., 85–86
Scotus, John Duns, 132, 138
sin, original, 131, 135–36
    personal, 146–47
Smith, Huston, 147
Sober, Elliott, 81–83
Society, Whiteheadian, xv, xvi, 7–10, 11, 21–25, 27, 37–42, 44, 53, 55, 71–74, 76–89, 92, 94–101, 103, 105–9, 112, 115–16, 120, 123, 129, 134, 141–43, 148, 151, 156, 158–61, 164, 167, 173, 179–82
    structured society, 11, 24, 27, 39–41, 71–72, 80, 86–89, 107, 120–21, 140, 142–43, 146, 148, 151, 159–60, 164–66, 171–75, 182
    common element of form, xvi, 27, 38, 40–42, 44, 53, 71, 76, 77–79, 94–95, 98, 103, 109, 115, 117, 120–21, 140–42, 148, 158–60, 164–65, 167, 179
    conceptual reversion, 160
soul/mind vs. body, 65, 86–89, 127, 139–53, 172 (*see also* consciousness)
Southgate, Christopher, 19–20
Space. *See* field of activity: structured
Spinoza, Benedict, 63
spirit/matter, 65–66, 158
    absolute spirit, 92
spontaneity, 6, 26, 70, 80, 82, 95
Stoeger, William, 51
subjectivity/objectivity, 22, 66, 72, 91–109, 123, 146 (*see also* actual entity)
    self, 64–65, 178
    subject of experience, 7–12, 21–22, 36, 38–42, 71, 76, 80, 83, 85, 93, 99, 101, 103, 107, 109, 117, 119, 138, 140–43, 148, 151, 158, 162, 167, 173, 177–78
    subject/superject, 26, 106, 148, 151, 158, 162
substance/substantial form, 23, 56, 61, 63–64, 72, 85–86, 88, 96, 99–100, 101, 141, 148, 163, 179
Suchocki, Marjorie, 151
suffering
    animal, 19–20, 34–35, 48, 131
    divine, 48–49
    human, 19–20, 34–35, 48, 131

system/systems theory, 55–56, 62, 71–74, 91–92, 99–100, 102–7, 118–19, 121, 123–24, 167, 179–82
   adaptive unities, 78–81, 84, 88
   lower-level vs. higher-level, 120
   open-ended, 70, 74, 75–89, 109, 111–24
   self-organizing, 31–45
   self-referential, 103, 106
   self-unifying, xii, 73–89
   thermodynamic, 32–33
   totalizing, xi, 67, 69, 92–93
system and environment, 103–6, 107–8, 116, 118–20, 179

Teilhard de Chardin, Pierre, 44, 157–58
theodicy, 20
time vs. eternity, 22
Trinity, doctrine of, xvi, 10, 16–17, 20–22, 44, 48, 56–57, 112, 127–38, 143–44, 167, 170, 173–74, 177
truth/objective vs. subjective, 53, 68–69, 93
   as relational and communitarian, 69

uniformity vs. regularity, 6, 9
union vs. unity of subject(s) of experience, 128

value, 15–29
Van Till, Howard, 43

Ward, Keith, 139–53
Weber, Michel, 95, 109
Whitehead, Alfred North, xv n. 12, xvi, 4, 7–10, 15–16, 21–29, 32, 34, 36–42, 44, 47, 52–58, 73–74, 76–89, 91–109, 112, 115–17, 119–21, 123, 128–30, 134, 140–48, 151, 156–60, 164–65, 167, 171–72, 173, 179–82
   Adventures of Ideas, 85, 93, 96
   Process and Reality, 7 n. 14, 17, 25, 31, 36–39, 41–42, 53–55, 70–71, 73, 78–79, 85, 93–94, 96, 106, 116, 120, 141–44, 147–48
   Science and the Modern World, 15, 18
whole vs. parts, 141, 145–46, 178, 182–83
Wiehl, Reiner, 91–93, 101
will of majority vs. general will, 114–15
Wilson, David Sloan, 75–83, 85, 88–89
world religions, 174–75, 181–82

Zycinski, Jozef, 31–36, 44

www.ingramcontent.com/pod-product-compliance
Lightning Source LLC
Chambersburg PA
CBHW071912290426
44110CB00013B/1359